Taking Science to the People

A Communication Primer for Scientists and Engineers

Edited by Carolyn Johnsen

University of Nebraska Press
Lincoln and London

Library of Congress Cataloging-in-Publication Data
Taking science to the people: a communication primer
for scientists and engineers / edited by Carolyn Johnsen.
p. cm.
Includes bibliographical references.
ISBN 978-0-8032-2052-2 (pbk.: alk. paper)
1. Communication in science. I. Johnsen, Carolyn, 1944–
Q223.T35   2010
501′.4 — dc22
2010009017

Set in Quadraat and Meta.
Designed by Nathan Putens.

Taking Science to the People

Contents

Acknowledgments

Many thanks to the authors for their thoughtful contributions to this book; to UNL Vice Chancellor for Research Prem Paul for suggesting this project; and to Dr. Paul and my dean, Dr. Will Norton Jr., for supporting the project financially. Thanks also to Diandra Leslie-Pelecky, formerly on the faculty of the UNL Department of Physics, for taking the lead in organizing the 2007 conference and for her untiring commitment to communicating science to the public. Thanks to John Janovy Jr., a distinguished UNL professor of biology, for responding to the essays in the book's afterword. And thanks to the editors at the University of Nebraska Press for their advice and patience in seeing this book through to print.

Carolyn Johnsen is on the faculty of the University of Nebraska–Lincoln College of Journalism and Mass Communications, where she teaches science writing and reporting classes. For ten years, she reported on the environment and agriculture for the Nebraska Public Radio Network. Her stories were also broadcast on National Public Radio, the BBC, and Monitor Radio. Her work has received national and regional honors, including awards for writing and for environmental, investigative, feature, and documentary reporting. She is the author of *Raising a Stink: The Struggle over Factory Hog Farms in Nebraska* (University of Nebraska Press 2003). Johnsen has also been a Fulbright teacher in England. She has a bachelor's degree in education and a master's in English — both from UNL. She joined the UNL journalism faculty in 2004.

Introduction

Carolyn Johnsen

In July 2008 Rush Limbaugh, the conservative talk-show host, called the lead scientist at NASA's Goddard Space Institute "an idiot."

The epithet fit comfortably in the context of Limbaugh's daily rants against liberals, environmentalists, Barack Obama, and what Limbaugh has called the global warming "hoax." More than six hundred radio stations nationwide broadcast Limbaugh's show for three hours every day. So Limbaugh's opinion of James Hansen and his efforts to inform the public on the science of global warming reached hundreds of thousands of listeners. It would be repeated in coffee shops, subway stations, and offices nationwide.

On the other hand, Hansen's quiet defense of his science was carried, if at all, in ten-second sound bites on radio and TV programs that infrequently covered science news or in an inch or two of print in wire stories in the nation's newspapers.

In fact the Bush administration had tried to silence Hansen, who told Andrew Revkin of the *New York Times*: "In my 30-some years of experience in government, I've never seen control to the degree that it's occurring now. It's just very harmful to the way a democracy works. We have to inform the public if they're going to make the right decisions and influence policymakers."[1]

Protecting our democracy may be the most important reason for scientists and engineers to explain their work clearly to non-experts—whether to the press, the public, or policymakers.

Important public-policy debates on topics as diverse as global warming, stem-cell research, autism, health care, biogenetics, energy, and food safety call for the expert insight of scientists and engineers. Timely, accessible information from these experts can encourage policymakers to consider evidence along with ideology while making decisions. In fact, ideology untempered by empirical evidence can too easily lead to misguided policy related to human health and even to the health of the planet.

The changing role of the media also places a duty on scientists and engineers to provide expertise and clarity in policy debates related to science. In a limited and sometimes uneasy partnership with scientists, journalists have traditionally translated scientific and technical information for the public. But both print and broadcast media are cutting their coverage of science news, leaving a void of information at a time when we need it most.

Boyce Rensberger is the former director of MIT's Knight Science Journalism Fellowships and a contributor to this book. In Harvard's *Nieman Reports*, Rensberger wrote, "The impacts of science, including technology, and its effects on individuals and on society, are becoming more powerful and less predictable. It is more important than ever that the public be informed of what's happening in science."[2]

And yet traditional news media, which would typically be the conduit for this information, are showing less commitment to this role. An analysis from the Project for Excellence in Journalism, titled "The State of the News Media 2008," cites a study by Christine Russell of Harvard's Shorenstein Center that "estimates that of the 95 newspapers that published special science sections in the 1980s, only about 35 still do so today."[3]

The report also says that in watching five hours of cable news in 2008, a viewer would have seen at least twenty-six minutes on crime, ten minutes on celebrity and entertainment, and less than two and a half minutes on science, technology, and the environment.

The proliferation of Web sites and blogs dedicated to science offer one way for scientists and engineers to pick up the slack. Motivation to do even more comes from a major funding agency.

Many researchers rely upon National Science Foundation (NSF) grants that pay for important elements of research, such as laboratory space, equipment, graduate students, and travel. The NSF now requires researchers applying for grants to include plans for reaching beyond the laboratory to explain their work. Leslie Fink of the NSF elaborates on this obligation in her essay, in addition to giving advice on how to meet that obligation.

The NSF, in turn, feels pressure from Congress to expand the public's access to information about science. In 2007 Congress passed the America COMPETES Act, which urged the NSF to do more to teach science graduate students how to communicate more clearly about their work to "nonscientist audiences."

The Act arose from a practical need for policymakers to have clear information on how researchers spend the tax money that supports them. But the public has a stake in this process as well.

In the NSF's periodic reports on public opinion about science and technology, the authors always observe that good citizenship relies, in part, on a knowledge of science: "Knowing how science works—how ideas are investigated and either accepted or rejected—can help people evaluate the validity of various claims they encounter in daily life."[4] Obvious examples include the competing claims of manufacturers of pain remedies and children's car seats.

But when politicians sort out competing points of view on such topics as food safety or coal mining's effects on the environment, their decisions affect us all. So researchers who build the foundations of science and technology have a critical role to play in bringing clarity to the discourse on both personal practice and public policy.

In his inaugural address, President Barack Obama vowed to "restore science to its rightful place."

The rightful place of science in a democracy is at the center of policymaking on many of the most pressing issues of our time.

In an essay for the *New York Times*, science writer Dennis Overbye pointed out the similarities between the values of science and the values of a democracy: "honesty, doubt, respect for evidence, openness, accountability and tolerance and indeed hunger for opposing points of view."[5]

These values—which echo those that drive good journalism—offer a further rationale for researchers in a democracy to communicate more clearly about their work.

Taking Science to the People is primarily for scientists and engineers who acknowledge these opportunities and obligations and who want to improve their communication skills. The essays published here should also persuade some skeptics to polish their communication skills and to provide the means for their graduate students to do so. Accordingly, the authors offer both the rationale and some tools for communicating about science and technology to nonexperts.

This book grew out of a conference with the cumbersome title "Communicating Science to Broader Audiences" held at the University of Nebraska–Lincoln in 2007. Speakers from the NSF and from university information offices, and journalists and "popularizers" of science who are, themselves, scientists, offered compelling evidence of the need for researchers to take on this outreach role.

The conference drew more than a hundred people, primarily from university public-information offices nationwide. Only one reporter attended—a troubling but unsurprising fact, given current trends in news coverage. But about one-third of the attendees were scientists or science graduate students who recognized the need to communicate about their work to the public.

In an effort to reach a wider audience of scientists and engineers, this book picks up the thread spun out at the 2007 conference. To that end, several speakers submitted chapters reflecting and expanding upon their comments at the conference. They include Leslie Fink, David Ehrenstein, Sidney Perkowitz, Stacey Pasco, Boyce Rensberger, and Margaret Wertheim. Some of those authors are either scientists themselves or have received graduate education in the sciences.

Three other authors—Georgia Tech science writer Abbey Vogel, journalist Warren Leary, and Gene Whitney, a government scientist—did not speak at the conference but were invited to contribute because of the perspective they could offer from their own experiences.

I hope scientists, engineers, and graduate students in the sciences and engineering will read this book and find the authors' insights and advice convincing and useful.

Although some of the authors have abandoned full-time scientific endeavors to write about science, none of them suggests that researchers must abandon either their work or their specialized language. Instead these writers urge researchers to become equally fluent in the plain English needed to communicate about their work to the public and policymakers.

While the news media sort out the future of science news and of journalism itself, this book may also remind journalists of the critical role they have played in effectively communicating to the public about science and technology.

Indeed, because journalists still offer a common route for information about science to reach the public, scientists can benefit from learning a thing or two about how journalists do their work. Several essays offer that information.

I add one thought not covered by any of the authors in this book: If scientists and engineers are to spread the word about their work to nonexperts (people other than peers), institutions who employ scientists — primarily universities — should develop a system of incentives and rewards for that effort.

Too often, scientists who "popularize" their work are rewarded not with praise but with their peers' scorn or indifference. One exception is the annual AAAS Award for Public Understanding of Science and Technology, recognizing "scientists and engineers who make outstanding contributions to the 'popularization of science.'"[6] To further efforts to communicate to the public, the AAAS even offers "Communicating Science: Tools for Scientists and Engineers."[7]

W. Wayt Gibbs, a contributing editor at *Scientific American*, told the 2007 UNL conference, "Most scientists see no reward for this kind of work. Until that's part of the job expectations, they're reluctant to do it." Gibbs added that the lack of reward for scientists who tell the public about their research makes such an effort seem more of a charitable activity than a professional responsibility.

Writing in the journal *Science Communication*, Michael Weigold of the University of Florida explained why scientists resist communicating with the public: "Fellow scientists may look down on colleagues who go public, believing that science is best shared through peer-reviewed publications. Scientists may also believe that . . . scientists should be humble and dedicated to their work, that scientists should have neither the time nor the inclination to blow their own trumpets."[8]

In contrast, science journalists have many opportunities to receive recognition for excelling in their work. Here are three examples of annual awards: The Society of Environmental Journalists gives cash awards to reporters for the best environment coverage aired, printed, or posted;[9] the National Association of Science Writers, Inc., gives the Science in Society Award to outstanding science journalism;[10] the Metcalf Institute for Marine and Environmental Reporting at the University of Rhode Island awards the $75,000 Grantham Prize to honor outstanding reporting on the environment.[11]

Some of the most prestigious awards for science journalism come from scientists themselves — from the American Association for the Advancement of Science and the National Academies of Science.[12]

Although the role and the very shape of news media are in flux, citizens in this democratic society still need information about science and technology. The authors collected in this book urge scientists and engineers to do their part to fill that need.

Margaret Wertheim, a distinguished science journalist educated in physics, mathematics, and computer science, issues this call to action: "It is time to get off our high horses and go out to the people."

NOTES

1 "A Conversation with Dr. James Hansen," Erik Olsen, producer, *New York Times*, http://video.nytimes.com/video/2006/01/28/science/1194 817097774/a-conversation-with-dr-james-hansen.html.

2 Boyce Rensberger, "Reporting Science Means Looking for Cautionary Signals," *Nieman Reports* (Fall 2002), http://www.nieman.harvard.edu.

3 "The State of the News Media 2008," Project for Excellence in Journalism, Pew Research Center, http://pewresearch.org.

4 The NSF's "Science and Engineering Indicators" can be found at http://www.nsf.gov/statistics/seind. The cited material was taken from chapter 7 of the 2006 survey.

5 Dennis Overbye, "Elevating Science and Elevating Democracy," *New York Times*, January 27, 2009.

6 AAAS Award for Public Understanding of Science and Technology, American Association for the Advancement of Science, http://www.aaas.org/aboutaaas/awards.

7 "Communicating Science: Tools for Scientists and Engineers," American Association for the Advancement of Science, http://communicatingscience.aaas.org.

8 Michael Weigold, "Communicating Science: A Review of the Literature," *Science Communication* 23, no. 2 (December 2001): 164–93, http://scx.sagepub.com/cgi/content/abstract/23/2/164.

9 SEJ Awards for Reporting on the Environment, Society of Environmental Journalists, sej.org.

10 National Association of Science Writers, http://www.nasw.org/awards/society.htm.

11 Grantham Prize, http://www.granthamprize.org.

12 AAAS Kavli Science Journalism Award, http://www.aaas.org/aboutaaas/awards/sja. For science journalism awards, such as the Keck Foundation Award, given by the National Academies of Science, see nas.org.

Taking Science to the People

Leslie Fink is a science communicator at the National Science Foundation. For twenty-five years, she has been involved in communication programs for federal research agencies in the Washington DC area.

Fink established and led the communications office of the Human Genome Project at the National Institutes of Health. She later led communications for the National Institute of Allergy and Infectious Diseases, the NIH component in charge of research on HIV, global infectious disease, and biodefense.

Recently, Fink has been producing multimedia Web pieces and exploring opportunities to include science and engineering themes in popular-culture venues, especially in movies and on TV.

She holds a bachelor's degree in biology, carried out cancer research at the University of Wisconsin–Madison, and completed the graduate program in science communication at the University of California at Santa Cruz.

In this opening chapter, Leslie Fink explains the obligations that federal law and policy set for researchers to communicate about their work to nonexperts. She also offers methods that researchers can use to avoid "Tower of Babel" consequences.

1

"The Difficulty of Nubbing Together a Regurgitative Purwell and a Superaminative Wennel Sprocket"

Leslie Fink

> We've arranged a global civilization in which the most crucial
> elements . . . profoundly depend on science and technology. We
> have also arranged things so that no one understands science and
> technology. This is a prescription for disaster. We might get away
> with it for a while, but sooner or later this combustible mixture of
> ignorance and power will blow up in our faces.
> CARL SAGAN, *The Demon-Haunted World*

Perhaps the best-known story of the untoward consequences
of bad communication is the biblical account of the Tower of
Babel, in which God is said to have created the world's different languages in order to prevent the tower's builders from
understanding one another. As intended, the babble that erupted
among them brought the project to a halt.

In modern times, the Tower of Babel metaphor may aptly apply to communication between scientists and the public, with similar consequences to the support of research through funding and social acceptance.

Like most specialists, scientists have refined a way of communicating that operates effectively in the halls of academia and in professional societies but falls short in popular parlance. This chapter's title — "The Difficulty of Nubbing Together a Regurgitative Purwell and a Superaminative Wennel Sprocket" — provides a humorous but very real example.[1]

The history of science tells us that the communication difficulties between scientists and nonscientists are a relatively recent occurrence — one that may be related to the shift in funding sources that occurred in the last sixty years.

Until at least the mid-nineteenth century, theorizing, research, and exploration were carried out by "men of science" — mostly savants supported by wealthy patrons or private foundations.

In fact support of science as a public investment, particularly at universities, did not gain a foothold in the United States until after the end of World War II. Then, as the Cold War escalated, most federally funded research taking place at universities was supported by military contracts, not by the system of grants awarded by peer review that is common today.

In 1954, for example, the Department of Defense and the Atomic Energy Commission supported 96 percent of academic research outside of medical and agricultural studies. From "V-J Day to Sputnik," those funds were concentrated at a few of the nation's elite campuses and supported very directed, applied research on weapons and other military technologies.[2] Working to outsmart real or perceived Soviet threats, academic scientists reasonably kept conversation about their projects to themselves.

When the Cold War ended, emphasis in science-funding policy in the United States shifted toward an increasingly diverse portfolio of peacetime pursuits to improve quality of life and economic growth. Today the public is both benefactor and beneficiary of those policies, as applications of federally funded science-and-engineering research have made their way into nearly every aspect of American life.

Still, most tenured research faculty in university labs today were Cold War scientists themselves (or were trained by someone who was) who instilled the culture of the period in their students. In contrast to the mum culture of the Cold War years, scientists have a responsibility now more than ever to participate in dialogues about their work directly with citizens who will, in the voting booth, ultimately decide its intellectual, practical, or even moral value.

Citizens are being called upon to make decisions about increasingly complex scientific and technological issues, such as climate change, stem-cell research and its applications, energy policy, green technologies, evolution, genetically modified foods, privacy issues related to surveillance, computer and medical technologies, space exploration, defense technologies, education, and end-of-life decisions, to name a few.

THE PUBLIC AS BENEFACTOR AND BENEFICIARY

> The most important things happening in the world today won't make tomorrow's front page. . . . They'll be happening in laboratories — out of sight, inscrutable and unhyped until the very moment they change life as we know it.
> JOEL ACHENBACH

In a 2008 *Washington Post* article, journalist Joel Achenbach went on to say, "We vaguely understand that this stuff is changing our lives, but we feel as though it's all out of our control. We're just hanging on tight, like Kirk and Spock when the Enterprise starts vibrating at Warp 8."[3]

Indeed years of public surveys conclude most people know relatively little about science and technology. A National Science Foundation (NSF) survey showed that nearly half of the U.S. adults questioned did not know how long it takes Earth to orbit the sun. About half did not know that electrons are smaller than atoms, and only one-third knew that the universe started with the Big Bang.[4]

Nevertheless, the same surveys show that all segments of the American public overwhelmingly support scientific research and the federal government's funding of it. About 80 percent of survey respondents said the federal government should support research "even if it brings no immediate benefits."

By 2006 the federal government provided the majority of funding for academic research and development—63 percent. Six agencies supplied about 95 percent of the $25 billion spent in 2007. NSF is the lead federal agency for funding academic research in the physical sciences, mathematics, the computer sciences, and earth, atmospheric, and ocean sciences.[5]

By law, NSF and other federal agencies now establish periodic strategic plans and annual mechanisms to report progress to Congress and the administration's Office of Management and Budget.

One such mechanism is the requirement that researchers who receive federal funds regularly notify their agency program managers, who are themselves scientifically trained, of their progress. Such transmittals come in the form of published research papers, annual progress reports, and other technical documents. NSF also asks for brief lay summaries, called "Highlights," from its investigators.

First and foremost, these would-be simple statements fulfill requirements of the Government Performance and Results Act (GPRA, pronounced "gippra") of 1993. A key GPRA objective is to "improve the confidence of the American people in the capability of the Federal Government, by systematically holding Federal agencies accountable for achieving program results."[6]

Highlights help NSF leadership account for the agency's management of funds to congressional appropriators. Highlights also serve to make a persuasive case for increases in future funding to budget officials in the executive branch, who parse the scarce tax dollars in the very competitive budget the president submits to Congress each fiscal year.

Besides contributing to bureaucratic reporting requirements, well-thought-out and well-written research Highlights can serve a number of other useful purposes. They can be especially valuable in articulating to stakeholders important research problems or knowledge gaps and in bringing distinction to institutions trying to solve them.

Scientists who can present their research "Highlight-style" are assets in communities seeking to parlay local intellectual talent into better schooling or business and economic benefits and better healthcare, for example, or simply a more enriched life for citizens.

Finally, those scientists are blue-chip commodities on the "Good Will Exchange" when the inevitable and widely chronicled misadventure threatens an institution's reputation.

Scientists can practice these short, clear explanations of their work by developing a so-called elevator speech — a good way to organize information about their research and why it's important.[7] Succinct elevator speeches can hone scientists' skills in framing highly technical work in words that have meaning to all — a benefit not just for conveying the significance or importance of individual research projects to the public but also to management and for garnering new resources.

NSF has partnered with the Center for Public Engagement at the American Association for the Advancement of Science to provide online resources including Webinars, how-to tips for media interviews, strategies for identifying public outreach opportunities, and more to help scientists and engineers develop public communication skills.[8]

In August 2007 President George W. Bush signed the America COMPETES Act into law. Included as a "Sense of Congress," the law says that NSF should "train graduate students in the communication of the substance and importance of their research to nonscientist audiences." It directs NSF to report to Congress within three years the details of those training programs.[9]

The mandate originated from a bill called the "Scientific Communications Act of 2007," which Doris Matsui (D-CA) and Bart Gordon (D-TN) introduced in the House of Representatives "to help bridge the communications gap between scientists and the rest of us." Matsui said, "If scientists can't tell the rest of us what they've discovered, we are not fully realizing the benefits of our investment in scientific research."[10]

Ample anecdotes indicate that Gen-X graduate students are not only willing and enthusiastic to talk about their work with nonscientists, many are also quite good at it. Formal training in communication can enhance and reinforce those skills.

In a letter to the journal *Science*, a group of scientists and communications experts at Cornell University reported on a course they designed to teach science communication to graduate students. Their goal was to improve students' abilities "to discuss our research with both the general public and the professionals writing and reporting on science in the media."

The authors made three suggestions. First, they said, involve people from multiple fields, especially those from the campus media-relations office, but also other scientists experienced in communicating with the public as well as journalists themselves.

The authors also recommended visiting a newspaper or a radio or television station and sitting in on editorial meetings in which editors and journalists pitch stories. That way, scientists can learn which findings are considered newsworthy and gain a better understanding of what journalists need when preparing a story.

Finally, the Cornell group suggested graduate students get hands-on experience by writing news releases, conducting interviews, being interviewed, and taking advantage of other opportunities to communicate with nonspecialists.

The letter concluded, "Starting public communication training at the graduate level will increase the frequency and confidence with which scientists communicate, with positive feedback for both science and public understanding."[11]

THE SCIENTIST IN THE PUBLIC SQUARE

> All mankind is divided into three classes: those that are immovable, those that are movable, and those that move.
> BENJAMIN FRANKLIN

Even technical journals have begun to express the point of view that increased communication between scientists and the public is a life-or-death matter for the research enterprise. Alan Leshner, chief executive officer of the American Association for the Advancement of Science, which publishes the journal *Science*, likens the current relationship between science and society to the Dickensian best of times, worst of times.

Alongside unprecedented advances in science and technology, society is "exhibiting increased disaffection," Leshner says, fostered by cases of data fraud and financial conflicts of interest. Worse, public skepticism and concern are increasingly aimed at scientific issues that appear to conflict with basic human values, religious beliefs, or political or economic agendas.

"The ensuing tension," Leshner says, "threatens to compromise the ability of the scientific enterprise to serve its broad societal mission and may weaken social support for science."

Leshner acknowledges that encouraging graduate students to communicate better directly with the public may come at some risk: "Many young colleagues are enthusiastic about discussing their work with the public, but they are also under tremendous pressure to stick to the bench, secure hard-to-get research grants, and publish rapidly in high-quality journals. Many even feel that the culture of science actively discourages them from becoming involved in public outreach, because it would somehow be bad for their careers."

In the end, public understanding of scientific facts is not sufficient because even (or especially) people who have command of the science may still have trouble embracing it in a societal context.

Leshner says, "We must have a genuine dialogue with our fellow citizens about how we can approach their concerns and what specific scientific findings mean."[12]

Former NSF director and White House science adviser Neal Lane defined the "civic scientist" as "one who uses his or her special scientific knowledge and skills to influence policy and inform the public." Lane considered Benjamin Franklin to be the model civic scientist owing to his command of three important qualities: wisdom, science, and communication.

Franklin was indeed a wise man—early to bed and early to rise. Lane attributes Franklin's wisdom, in part, to his older age compared with his Revolution-era contemporaries. Franklin was also a scientist who was elected a fellow of the Royal Society of London and of France's Royal Academy of Sciences. For his discoveries in electricity, Franklin received recognition equal to today's Nobel Prize. Both scientists and the public read his book, *Experiments and Observations on Electricity*.

Franklin's skill as a public communicator was perhaps most evident when, under the pseudonym Richard Saunders, he published *Poor Richard's Almanack*. From 1732 to 1758, Franklin filled yearly almanac pages with sprightly accounts of weather, astronomy, and even astrology, as well as poetry, math problems, aphorisms, proverbs, and other musings. The *Almanack* was reportedly the second-best-selling book in the colonies behind the Bible.

According to Lane, Franklin "would not be timid about convening town meetings where community leaders and other citizens could candidly discuss with scientists the moral, ethical, and practical implications of cloning, stem cell research, genetically modified crops and foods, nanotechnology, nuclear energy, missile defense, and so forth. And he would encourage scientists to listen as well as talk. No doubt Franklin, who taught by example nearly everywhere he went, would ask scientists of all disciplines to become more personally involved in their communities."[13]

Along those lines, NSF now requires funding applicants to address two equally important criteria in their proposals. The first, *intellectual merit*, addresses, among other things, how important the proposed activity is to advancing knowledge and understanding within its own field or across different fields. The second, *broader impacts*, addresses how well the activity advances discovery and understanding while promoting teaching, training, and learning; broadens the participation of underrepresented groups; and describes what the benefits to society may be.

Today, it is NSF policy to "return without review proposals that do not separately address both merit review criteria."[14]

PUBLIC UNDERSTANDING OF SCIENTISTS

Everything should be made as simple as possible, but not simpler.
ALBERT EINSTEIN

Despite the difficulty scientists may think they have communicating with nonscientists, surveys show that researchers enjoy an admired position of prestige and credibility among the public.

A survey by Research!America reported that scientists topped a list of admired professions with 57 percent of respondents saying scientists had "very great prestige." In contrast, only 15 percent said journalists did. Thirty percent said members of Congress did.[15]

Communicating with an interested, intelligent (but not expert) public requires the same considerations as communicating with colleagues in a different, but equally rigorous, profession. Most members of the public certainly will not have the same knowledge an expert does. But they can and do understand the information when that knowledge is communicated effectively.

Making understanding possible requires the expert and the nonexpert to connect in a shared, neutral space in which neither party is in control, but in which both parties stretch beyond their comfort levels. It is never a matter of "dumbing down." The goal is to make understanding happen by taking into account different experiences and points of reference in the communication process.

To accomplish this goal, scientists often rely on journalists, who have long been the primary purveyors of science to the public. The Internet has modified that role, but scientists still have to talk to journalists. Scientists and journalists, however, are trained to present information differently. Scientists begin with ample history and background followed by the facts and their context—what the finding means in the bigger picture. Journalists and members of the public, on the other hand, consume information in the opposite order, with the most important questions being, What is the discovery? and What does it mean? The details will be interesting to some, but in journalism and other public communication, the news comes first. Other authors in this book explain in more detail the benefits that accrue when scientists and journalists communicate with each other.

COMMUNICATING SCIENCE IN THE DIGITAL
AGE AND OTHER OPPORTUNITIES

> The clashing point of two subjects, two disciplines, two cultures —
> of two galaxies, as far as that goes — ought to produce creative
> chances.
>
> C. P. SNOW

The federally funded research enterprise in the United States has become an endeavor carried out by a diverse group of people in the sunshine of public scrutiny and accountability. Now we urge scientists to leave their comfort zones to communicate with the public at a time when communication technologies have never been so daunting.

The handful of major national newspapers that once reported science has largely given way to countless cable programs, Internet news sites, blogs, personal-device downloads, wireless transmissions, and the like.

The endless media formats now available, combined with the pervasive role of science and technology in everyday life, give scientists and their research institutions unprecedented opportunities to communicate directly with the public.

Today, popular culture, including art, music, sports, television, and movies — even video games — are "the current vernacular" and offer myriad opportunities to engage the public in interesting and relevant ways.

Opportunities lie elsewhere, as well, in an age where academics are encouraged to collaborate across disciplines.

For example, many universities support humanities programs that can enrich the presentation of science and engineering when invited to partner in broader-impact activities. A research theme may lend itself just as easily to a dramatic film or play, musical performance, or art exhibit as to the now-traditional Web site.

Experience tells us that students and faculty in creative, non-science departments are eager to take on intriguing technical topics, particularly those with complex societal implications. Experience also tells us the "two cultures" are not as different as they may once have been.

Creative chances abound.

NOTES

1 J. H. Quick, "The Turbo-Encabulator in Industry," *Student's Quarterly Journal*, Institution of Electrical Engineers, London, 1944. A video version of the complete article can be found at http://www.youtube.com/watch?v=PtuqjFf7-N4&feature=related/.

2 Roger L. Geiger, "Science, Universities, and National Defense, 1945–1970," *Osiris*, 2nd ser., 7 (1992): 26–48.

3 Joel Achenbach, "It's heading right at us, but we never see it coming," *Washington Post*, April 13, 2008, "Outlook" section.

4 National Science Foundation, "Science and Technology: Public Attitudes and Understanding," in *Science and Engineering Indicators 2008*, 2: appendix table 7-5.

5 National Science Foundation, "Academic Research and Development," in *Science and Engineering Indicators 2008*, 1: chapter 5.

6 *Government Performance and Results Act of 1993*, Public Law 103-62, 103 Cong., 1st sess. (August 3, 1993), Sec. 2(b)(1).

7 An "elevator speech" is a way to practice briefly and clearly explaining to nonexperts what you do. Imagine yourself with one or more strangers on an elevator. None of them is an expert in your field. They may include people who finance your research, lawmakers considering withdrawing funding from your lab, your department chair, who has his own narrow specialty, or some of the 102 freshmen in your survey class. You have the time it takes to go from the first to the fourth floor to explain to your captive audience what you do and why it's important to them. For more on the elevator speech, see Christian Daughton, "Emerging Pollutants, and Communicating the Science of Environmental Chemistry and Mass Spectrometry: Pharmaceuticals in the Environment," *Journal of the American Society of Mass Spectrometry* 12, no. 10 (October 2001): 1067–76.

8 The NSF/AAAS advice on public communication skills for scientists is at http://communicatingscience.aaas.org/Pages/newmain.aspx/.

9 The America Creating Opportunities to Meaningfully Promote Excellence in Technology, Education, and Science Act, Public Law 110-69, Aug. 9, 2007, § 7035.

10 Doris Matsui, "House adopts Congresswoman Matsui's amendment to provide communications training for scientists," press release, May 2, 2007, http://matsui.house.gov/SupportingFiles/documents/070503_-_Science_Communications_amendment_passage.pdf.

11 Dana R. Warren et al., "Lessons from Science Communication Training," Science 316 (May 25, 2007): 1122. All the quotes from the Cornell group come from this source.

12 Alan I. Leshner, "Outreach Training Needed," Science 315 (January 12, 2007): 161. All preceding Leshner quotes come from this source.

13 Neal Lane, "Benjamin Franklin, Civic Scientist," Physics Today 56 (October 2003): 41–46. All preceding Lane quotes come from this source.

14 Rita R. Colwell, "Implementation of New Grant Proposal Guide Requirements Related to the Broader Impacts Criterion," Important Notice to Presidents of Universities and Colleges and Heads of Other National Science Foundation Grantee Organizations, notice no. 127, July 8, 2002.

15 Cristine P. Brown et al., "Report: Helping Researchers Make the Case for Science," Science Communication 25, no. 3 (March 1, 2004): 294–303.

Margaret Wertheim is an internationally noted science writer whose work focuses on the relations between science and the wider cultural landscape. She is the author of *Pythagoras' Trousers* — a history of the relationship between physics and religion, and *The Pearly Gates of Cyberspace: A History of Space from Dante to the Internet*.

Wertheim founded the Institute For Figuring, which is devoted to enhancing the public understanding of the aesthetic and poetic dimensions of science and mathematics.

Wertheim has a BS in pure and applied physics and a BA in mathematics and computer science. Her writing has appeared in the *New York Times*, the *Los Angeles Times*, *The Sciences*, *New Scientist*, the *Guardian*, *Wired*, and *Best American Science Writing* (2003). In 2004 she was the National Science Foundation's visiting journalist to Antarctica. In 2006 she won the Print Media Award from the American Institute of Biological Sciences for two articles published in *LA Weekly*, sister paper to the *Village Voice*.

In addressing those attending the 2007 science-writing conference at the University of Nebraska–Lincoln, Margaret Wertheim offered this advice: "Don't report science as something that happens in test tubes but as a deeply human activity that has happened throughout history." This advice reflects her own efforts to engage readers by telling the stories of science. In this essay, she argues that work remains to be done to reach a wider audience with those stories. Wertheim presents evidence showing that magazines intended to inform the general public about science are reaching only about half the possible audience.

2

Who Is Science Writing For?

Margaret Wertheim

We all know the dismal statistics: Our children's test scores on international assessments of math and science literacy are plummeting; the number of doctoral students in science and engineering is at a forty-year low; we are desperately short of science teachers; intelligent design is spreading like kudzu; and most of our citizens believe in ESP, angels, or alien abductions. There is much public hand wringing, and those of us who love science have good reason to worry.

The question we face is how to respond. As someone who has been writing about science for the general public for more than twenty years, I would like to suggest that some radical changes are called for in our strategies for communicating science.

The very verb we are dealing with points to the nub of the problem. Unlike speaking and writing, "communicating" supposes an active engagement on the part of an audience. For something to be communicated, it has to be not only transmitted but also received.

Yet in discussions about how to improve the public's understanding of science — of which there are an escalating number — it seems that only one side of this channel is addressed. We ask: How can we better transmit the findings of science? As a journalist, I wrestle with this daily and feel thrilled when I have managed to coin an elegant article on the ecology of a termite's gut or the mechanics of a spider's eye.

But there is another question that has, I think, been factored too little into our public discourse: Who is on the receiving end of our missives? In short, who are we writing for?

The primary public resources about science are popular science magazines. It is worth asking: Who buys them? Who reads them? The answers surprise many scientists — and many professional science communicators, too.

Eight top-selling science magazines — *Scientific American*, *Discover*, *Popular Science*, *Wired*, *Natural History*, *Science News*, *Astronomy*, and *Science* — collectively sell about 4.5 million copies a month. In all, they claim around seventeen million readers. In magazine-world parlance, a "reader" is someone who spends at least half an hour with an issue.

Reader numbers, which are quoted to attract advertisers, are notoriously optimistic, but let us give the benefit of the doubt here and say that seventeen million Americans are looking at some science magazine each month.

Who are they? In a nutshell, they are overwhelmingly well-educated men over forty in the upper socioeconomic brackets. I gathered the statistics as an exercise a few years ago; the figures available then were for 2002, but I very much doubt they have changed significantly in the intervening years.

These are the facts: In 2002 the median age of *Scientific American* subscribers was forty-nine; for readers, it was forty-six. The median age of *Discover* readers was forty-one; of *Popular Science* readers, forty-three; and of *Science News* subscribers, forty-nine. Of *Scientific American*'s subscribers, 87 percent were men and 13 percent were women. *Wired*'s subscribers were 85 percent male, 15 percent female; *Science News* subscribers were 72 percent male. A representative at *Popular Science*, by far the biggest selling, laughed when I asked for a gender breakdown and said I could safely assume the vast majority were men.

Of *Scientific American*'s subscribers, 85 percent had college degrees and 58 percent had graduate degrees. For *Science News* the figures were 78 percent and 46 percent. The median salary of subscribers was $87,600 for *Scientific American*, $90,800 for *Wired*, and $74,000 for *Natural History*.

Age also provides a window: Two-thirds of the audience of *Popular Science* and *Discover*—which together accounted for 2.5 million copies per month—were over thirty-five years old. For *Scientific American* and *Science News*, almost 80 percent of subscribers were over thirty-five. Of all subscribers, 22 percent were women. Most of the magazines did not break down their numbers by race.

According to the Census Bureau, the current U.S. population is 299 million. This means that 280 million people are not reading any science magazines.

Women, people under thirty-five, and those in the lower socio-economic brackets are barely being touched by the canonical channels of science communication.

Let me introduce, then, another set of facts. At the same time that I researched statistics on science publications, I also looked at women's magazines. Again I chose eight top sellers—*Vogue*, *Elle*, *Glamour*, *Cosmopolitan*, *Self*, *Redbook*, *In Style*, and *Good Housekeeping*.

In 2002 these magazines collectively sold 17.5 million copies a month. *Good Housekeeping* alone sold more copies than all eight science magazines combined (at 4.7 million a month), and none of the eight sold less than a million. With sales this huge, the women's magazine world does not always bother to collect reader statistics, but if we assume the number of readers per copy is similar to that claimed by the science publications, then close to seventy million people are reading a women's magazine each month.

It is perhaps a sad fact, but ineluctably a true one, that most women do not go near science magazines.

It seems to me that if we are serious about improving the public understanding of science, we have to start looking at where the public is — and if the mountain is not coming to us, then we must go to it.

It is for this reason that for many years, in my native Australia, I wrote columns about science for women's magazines such as *Vogue* and *Elle*. I considered this my missionary work.

Writing for the hairdo and hemlines set carries no cachet in the science world — and little in the science communication world, either — but I consider this some of the most difficult (and serious) work I have done. Believe me, it is harder to explain genetic engineering or Big Bang cosmology in the context of *Vogue* than in the infinitely more prestigious pages of the *Science* section of the *New York Times*, for which I also write. The most difficult work I have ever done, by a long shot, was the television science series I wrote that was aimed at teenage girls, made for ABC Australia.

As the flagship of alternative newspapers, the *LA Weekly* is known for its arts, culture, and political coverage; before me, they had never had a science writer, and it took me five years to convince them to let me do a science column.

It has been an honor and a pleasure and also a challenge. I have to assume my readers know nothing whatever about science and that even the most basic concepts must somehow be conveyed without seeming "teacherly." I have had the support of a wonderful editor, Tom Christie, who goes through my pieces with a fine-toothed comb, an open mind, and a naïf's questions.

I am sometimes staggered at the things Tom doesn't know, but I remind myself that if he doesn't know, then 99 percent of our readers won't, either. Yet ours is an educated and literate audience.

Scientists often think that science writers dumb down their work, skimping on details and eliding over subtle distinctions. But most science writers — myself included — also love to write long pieces that convey the intricacies of a subject. It is these stories that meet with the approval of scientists (whose approval we journalists naturally desire) and that generally win awards.

But the stark reality of our dollar-driven age is that print space is a precious commodity, and we are increasingly lucky to have any column inches for science. It is frustrating to have only 900 words — as I did for my columns in the Australian *Vogue*, or 1,200, as I do now in the *LA Weekly* — to describe something as complicated as bioremediation or the physics of freezing; but 900 words are better than no words, and in the context of improving the public's understanding of science, every one of them is precious.

Scientists, by training, are experts; the public, by default, are not — and the gap between these two domains is getting wider. It will not do to sit around and bemoan this fact and hope that one morning we will wake up and find that everyone is reading *Science*, or even *Popular Science*. They will not.

We may not like the creationists, but there is one thing we could learn from them: the power and the value of grassroots proselytizing. In short, those of us who love science are called upon to be missionaries.

It is time to get off our high horses and go out to the people.

Gene Whitney, PhD, is currently head of the energy section of the Congressional Research Service. During the administration of President George W. Bush, Whitney was assistant director for environment at the Office of Science and Technology Policy (OSTP), Executive Office of the President, in Washington DC. In that role, he was responsible for earth sciences, earth observations, climate change, natural hazards and disasters, energy, water, environment, and natural resources.

He is cochair of the U.S. Group on Earth Observations and is OSTP principal to the U.S. Climate Change Science Program. He directed the Future of Land Imaging Interagency Working Group and serves on the Subcommittees on Disaster Reduction and Water Availability and Quality of the National Science and Technology Council.

Whitney, who received his doctorate in geology from the University of Illinois, has authored or coauthored numerous scientific papers and abstracts. He received a National Research Council Postdoctoral Fellowship at NASA/JPL and a Senior Postdoctoral Fellowship at the École Normale Supérieure, Paris. He has worked with the governments of China, Russia, Pakistan, Algeria, Bangladesh, and Japan on energy and mineral resource issues.

Gene Whitney has said that policymakers need scientific and technical guidance "more than they know." In an April 2008 talk at the University of Nebraska–Lincoln, Whitney told science faculty and students that "science almost never carries the day" in public policy decisions. "More times than I'd like to admit, science gets trumped . . . and you end up with a decision that makes no sense scientifically."

And yet, in the same talk, Whitney said the process of making public policy needs the input of both politicians and scientists: "If not for politicians, we'd never get any decisions made. Think of what the world would be like if scientists were in charge. When would you have enough data?"

In this essay, Whitney encourages scientists and engineers to offer their expertise to politicians and other policymakers and gives specific advice on how to do so most effectively.

3

Taking Your Science to the Capital

Gene Whitney

Policymaking has become increasingly technical over the last half century. The development of atomic weapons during World War II was followed by the space program in the 1960s and by rapid developments in physics, aerospace, earth and climate sciences, medicine and biotechnology, electronics and computing, and material sciences.

All these developments have been the subject of policy debates at many levels of government.

To make decisions, policymakers — from city councils to federal agencies — must grasp an incredibly broad and technical array of material. Congress and the president, for example, make policies about stem-cell research, nanotechnology, advanced weapon systems, pharmaceutical regulations, climate change, and energy options.

Many people who make public policy related to science are educated and equipped for the task, but virtually all of them rely on the advice of technical experts.

In the United States, decision makers include the president, members of Congress, governors, state legislators, county commissioners, city officials and hundreds of other government leaders.[1] In fact, decisions requiring scientific input are not in the hands of a single person or institution; many players acting in many venues over many years steer the ship of policy.

Although most decision makers try to be well informed, they don't have the time, the staff, or the resources to gather all the information they need to make well-informed decisions on technical or scientific issues. Sometimes a decision can be postponed, but more often decision makers must act on the available information, even if it isn't always the best information.

Policymakers become informed — or misinformed — about these issues through a variety of sources, including the media, personal study, staff, and expert briefings or testimony. However, many highly technical decisions are made with little expert input or with incomplete or bad information.

In short, across the country, at all levels of government, officials are making decisions every day because they must be made, although the decision makers would often benefit from additional expert input and counsel.

This situation provides an opportunity for scientists and engineers to become involved in the decision-making process.

WHO WILL DO IT?

Few scientists and engineers are interested in engaging directly in the political decision-making process. After all, they chose a career in science or technology because their aptitude, skills, and interests are in exploration, discovery, innovation, and the intellectual challenges of scientific or engineering processes. Furthermore, the institutions they work for may discourage them from dealing directly with the political sector.

Yet given the great need for accurate information on public-policy issues related to science and technology, institutions and professional societies should identify people who can provide that information, orient them to the political process, and provide resources and encouragement to get involved.

Some scientists and engineers, for example, lack either the necessary communication skills or the interest in understanding the political context in which policy is made. However, many are excellent communicators and have sufficient political savvy to lend their scientific or technical expertise to the policymaking process.

Scientists and engineers have a responsibility to inform national and global policies that require a sophisticated understanding of science and technology. Although the availability of this information does not mean decision makers will use it properly, individuals, institutions, and professional societies must find ways to enter the process to ensure that objective information about science and technology is available to political decision makers.

Know the Rules of Your Institution

Very few scientists and engineers are self-employed or scientific free agents. Virtually all work for government agencies, universities, nongovernmental organizations, or corporations that have rules dictating how their scientists and engineers can engage in the policy process and what information they can share.

Know your employer's policies and rules. Be sure to obey them, and keep your government-affairs office in the loop at every step.

Be Clear About Your Purpose — to Inform or to Persuade

If you find yourself in a politician's office without a clear reason for being there, you will be either ineffective or embarrassed or both.

Be clear to a congressperson, staffer, or political operative that you have come to accomplish something: You want to provide vital information on an issue, or you want to use your scientific or technical expertise to try to persuade the decision maker to favor a specific outcome.

If you want to *inform* a policy or decision, don't be naïve about political staffers' intelligence and level of knowledge. Although you have mastered the literature in your area of expertise, try to appreciate the volume and technical level of information that political staffers deal with every day. For this reason, try to gauge the level of technical expertise of the policymaker and staff before starting the conversation—which should be a dialogue, not a monologue. Ask questions to find the appropriate level of discussion:

Have they previously received a scientific or technical briefing on the topic?

What level of technical or scientific information do they need to make a decision?

What questions remain, and how might you help answer those questions?

Describe your background, specialty, and the interest you have in the issue; ask if this information will be useful.

Depending upon the responses to these queries, you may need to do some more homework before returning for a briefing. Don't be afraid to admit this. Ultimately, your assistance will be most valuable if it addresses gaps in understanding or helps to correct misinformation. Politicians don't always welcome voluntary input from scientists not involved in the policy process, so provide only the information that's needed.

If your intent on visiting a decision maker's office is to *persuade*, be aware of the rules, the potential payoffs, and the dangers.

Attempting to affect the outcome of a political decision through persuasion is lobbying. The U.S. Senate provides this information on lobbying:

> Lobbying is the practice of trying to persuade legislators to propose, pass, or defeat legislation or to change existing laws. A lobbyist may work for a group, organization, or industry, and presents information on legislative proposals to support his or her clients' interests.
>
> The Lobbying Disclosure Act of 1995 establishes criteria for determining when an organization or firm should register their employees as lobbyists. Lobbyists register with the Senate Office of Public Records.[2]

Most scientists and engineers would never consider themselves lobbyists. Yet some who begin working within the political process to inform decision makers become concerned about a decision's outcome and unintentionally begin to make recommendations favoring a certain outcome.

Be sensitive to the process and recognize when your thoughts and feelings are no longer totally objective and nonpartisan. Can you still present pros and cons on an issue with equal objectivity? Are you frustrated that a certain outcome isn't obvious to the decision maker?

If you lose objectivity on an issue, reevaluate whether you want to inform or to persuade, recognizing these as very different roles.

You can advocate for objective information without advocating for a specific outcome on an issue. In other words, you may simply want to persuade a policymaker to look at the data, to examine scientific findings, and to weigh the technical evidence before making a decision.

This advocacy for information doesn't require a policy position, but rather encourages and facilitates the use of available information to make informed policy decisions.

QUALITIES OR SKILLS A SCIENTIST OR ENGINEER NEEDS TO INFORM POLICY

1. Scientific or technical expertise

Expertise is often limited to a narrow specialty. For example, in order to evaluate the terms of a containment policy, decision makers may need to know how certain radionuclides are transported in groundwater. Experts with that specific knowledge are valuable to decision makers. However, specialists in atmospheric chemistry (for example) should limit their involvement to issues requiring that specialty.

2. Ability to communicate technical issues in plain English

Expertise has limited value if an expert can't communicate in terms that policymakers can understand. A politician may not know about the Clausius-Clapeyron equation, but may easily understand the importance of phase transitions in climate change if described in simple terms. An involved scientist or engineer has the responsibility to explain complex phenomena clearly when they are important to policy formulation.

Briefings for decision makers are different from the presentations that you typically make at professional conferences or symposia. In your briefing, tailor your language, length, structure, and materials to the needs of the policymaker.

3. Ability to communicate the uncertainty inherent in scientific information without sending the message that there is no "right answer"

Because so much science is probabilistic rather than deterministic, it is important to explain to policymakers that certain physical processes may have "likely" outcomes even though no single correct answer is apparent.

For example, when estimating the amount of petroleum in the unexplored territories of Alaska's North Slope, a geologist may express the estimates as a statistically significant value with associated uncertainties. A policymaker may say, "Just give me the answer," but one role of the scientist is to explain that scientific determinations aren't simply facts and often have uncertainty associated with them. In fact, the "answer" may change with additional data. The use of weather forecasting or some similar probabilistic analogy is often useful to explain the concept.

4. Ability to use scenarios to communicate effectively the possible consequences of decisions

In complex systems such as the natural interaction of atmosphere, ocean, lithosphere, and biosphere under conditions of changing climate, the result of an action may depend upon so many variables that the cause-effect relation can be described only in scenarios. Policymakers may be frustrated by the answer, "It depends."

Therefore, in attempting to inform policy involving complex natural or economic systems, be prepared to describe a handful of scenarios that reflect the nature and magnitude of uncertainty in the behavior of a stimulated or disturbed system. These scenarios may be based on system models, and a careful and general description of the model may help explain the possible outcomes. However, few policymakers want or need detailed explanations of these complex models; policymakers will be more receptive to an explanation of how the important variables are related in complex ways.

5. Recognition that scientific information might be used or misused by proponents of either side of an issue

One potential problem arising from the use of probabilistic scientific information or of complex scenarios is that the information might be used improperly to depict an unlikely outcome. For example, the probability of discovering a specific volume of oil in northern Alaska may have a statistical estimate with an associated range of probabilities. Proponents of development might select the probability indicating the highest volume of oil. Opponents might select the probability indicating the lowest volume of oil.

Without understanding the proper use of the numbers, each group may feel justified in choosing those values because they are "in the literature." Explain carefully when expressing outcomes as probabilities, and anticipate how such outcomes might be misused.

6. Ability to participate in a highly confrontational or contentious policy debate without losing sight of the objective scientific evidence

Politicians and policymakers care deeply about the issues they work on. This commitment often leads to conflict or disagreement. The strength of disagreement among advocates of opposite sides on a political issue may surprise you. In fact, confrontational political debate has caused some scientists to disengage from the process. If you're dedicated to the rigorous informing of policy on scientific and technical issues, understand that such confrontation isn't unusual and should (usually) not be taken personally. Thick skin and perseverance are important qualities in the political realm.

Scientists and engineers who are unfamiliar with the structure or workings of politics but who are qualified and willing to inform the policymaking process may need guidance on entering that process and functioning successfully. To be most effective and to avoid accusations of inappropriate lobbying or other negative motivations, always work through your institution's government-affairs office.

1. Identify the real decision makers

Although a president, senator, or member of Congress may make the final decisions, key staff members conduct most of the background investigations needed to arrive at a policy position. Avoid going directly to decision makers without first talking with their staffs, who are often highly influential in determining the direction a decision maker takes.

2. Understand how decision makers think

Only rarely are politicians and other decision makers also scientists. In fact, most decision makers in federal executive or legislative positions are lawyers. Lawyers are advocates. Their advocacy is essential to the policymaking process.

Lawyers and scientists use different methods in their work. Scientists must understand that lawyers aren't trained in inductive reasoning, as scientists are. Scientists collect and draw conclusions from data. If the data change, the conclusions may also change. Lawyers, however, use legal reasoning, which examines the facts in light of a set of rules, laws, or precedents. Scientists and lawyers looking at the same set of observations may not reach the same conclusions. Scientists venturing into the policy world must understand this difference so they aren't frustrated with the advocacy process.

3. Identify other scientific participants — know what they are thinking

Scientists continually discuss, debate, and challenge each other over scientific conclusions. Although many fundamental science precepts are considered fact, much scientific progress occurs via a process of testing and examination of data and conclusions. Interpretation of those factors may lead scientists to disagree.

Be sure to know which parts of your area of science are the objects of scientific consensus and which parts continue to be debated. Even establishing which areas of science have been "settled" is debatable, so you must be able to explain why some hypotheses continue to be debated. Policymakers looking for "the answer" will understand the uncertainty of scientific progress if properly explained; those policymakers, however, may also be frustrated by the lack of a definitive answer. If they can find an alternative conclusion in the scientific literature that better satisfies their political agenda, they may seize the opportunity to use it. Therefore, you must understand all sides of a scientific issue and be aware of scientists who propose alternative conclusions.

4. Anticipate the need for technical information; early information is best

Don't wait to offer technical information until the Senate is voting on a piece of policy legislation. The process of informing policy and decisions must start very early in the life of a policy discussion. In fact, you will sometimes be able to anticipate when an issue will rise to a level requiring political decisions. For example, early research in genomics led researchers to anticipate many of the ethical and legal issues that would eventually arise out of this powerful body of scientific work.

Objective, scientific and technical information is always helpful to policy decisions, regardless of the stage of debate; but to be most effective, provide that information as early in the process as possible.

5. Meet face to face rather than sending paper or e-mail

Scientific and technical information about a policy issue is relevant and helpful only if the intended recipient actually sees and understands it. Mailing publications from technical journals to a decision maker will have almost no effect. Sending white papers or e-mailed lists of recommended Web sites to a staffer may have some use but will most likely get lost in the flood of mail and e-mail typically received by a political office.

The most effective way to communicate with decision makers and their staffs is in face-to-face meetings, which are difficult to arrange. If you attempt unannounced meetings or knock on doors individually, you may find little success. So look for some help. Your institution's government-affairs office can arrange briefings and meetings with decision makers, who are typically more receptive to contacts from government agencies, universities, and nongovernmental organizations than to contacts from individuals. Take advantage of your institutional position and resources.

6. Establish technical credibility; stay in your lane

Scientists are valuable for their scientific expertise and generally not for expertise in public policy. A nuclear physicist isn't necessarily an expert in international nonproliferation negotiations. A wildlife biologist isn't necessarily an expert in the application of the Endangered Species Act. Be as well informed as possible about the political context for the issue you're interested in (see #7, below), but policymakers most need your guidance and insight on scientific or engineering matters.

7. Establish political credibility — know the context of the issue

The most valuable briefings are those in which scientists or engineers provide technical information to address a political question being asked by a decision maker. So beyond your technical knowledge, you must also know something about the broader context of the issue, both scientifically and politically.

Find out why a decision maker is seeking the information you can provide. What problem is being solved? What question is being answered? You may be able to speak for hours about your specialty, but a policymaker has specific questions, and you'll be most helpful if you focus on the issue at hand. A prebriefing conversation with the political office will help you identify the focus and may provide time to become better informed on the political issue being addressed.

8. Put yourself in the policymakers' shoes: What else are they worried about?

Political context for a technical issue may be very broad. For example, if the immediate problem is one of data security in federal agencies, an engineer may be asked to provide a briefing on hardware and software security vulnerabilities on the Internet, identity theft, and the security of international online banking. An engineer who can anticipate that the discussion may become more wide-ranging can prepare appropriately and increase the value of the briefing. It is impossible, of course, to anticipate every question, but some homework on the responsibilities and interests of the politician being briefed can be very helpful.

9. Offer to help; be a resource

Scientists and engineers with good communication skills can have an enormous impact on informing policy if they remain available for ongoing information and discussion. Carry business cards to signal to people that you're available. Punctuate each contact with the decision maker with an offer to provide more help.

Frequently, the first briefing on a technical issue produces more questions in a politician's mind. A follow-up discussion—whether in person, on the phone, or by e-mail—may provide valuable clarification or expand the range of the discussion. Sometimes, you may need to refer the decision maker to a colleague with more appropriate expertise.

10. Follow up

If policymakers ask for more information, send it promptly. If they ask for more briefings, try to make yourself available or engage a colleague to fill in. By providing follow-up clarification and information, you will have established an important relationship that will benefit you, the policymaker, and the nation.

11. Be persistent

Policy agendas have an ebb and flow that bring specific issues to the forefront at different times. If you have invested time and energy in briefing a policymaker on an issue and it hasn't been resolved or the policy hasn't been enacted, your work wasn't wasted. Many issues will surface again, and policymakers will need the same or updated information. The political process isn't linear; don't expect tidy outcomes for every effort. Some important policies require several attempts over a period of years to reach action and implementation. For example, the issue may remain but the political players change; or both the issue and the players stay the same, but the political landscape becomes more receptive to the policy.

Keep the scientific and technical information updated, and be prepared to present more briefings. Seize opportunities when they arise.

Scientists and engineers who provide scientific and technical information for decision makers at all levels of government provide a critical service to the nation and may find their efforts to be a satisfying aspect of their professions.

NOTES

1 The terms "decision maker," "policymaker," and "politician" are used interchangeably in this chapter to represent any official needing scientific or technical information to make a decision.

2 Requirements and guidelines for lobbyists can be found at http://lobbyingdisclosure.house.gov/.

Stacey Pasco manages the Mass Media Science & Engineering Fellows program at the American Association for the Advancement of Science (AAAS). She also manages the Minority Science Writers Internship at *Science* magazine (www.aaas.org/mswi), which provides opportunities for minority undergraduate journalism students to experience the rewards of science writing. Pasco has produced Web content for AOL and contributed to outreach components for the television series *Journey to Planet Earth: An Annual Report*.

The highly competitive Mass Media Science & Engineering Fellows program sponsored by AAAS places senior undergraduate, graduate, and postgraduate science, engineering, and mathematics students at media organizations nationwide to work as paid reporters, editors, researchers, and production assistants. The ten-week summer program is designed to hone the students' communication skills — particularly those needed for communicating with the public. In this essay, Stacey Pasco describes the goals of that program and its effects on the students who participate and on the newsrooms where they work.

4

Building a Better Science Communicator

Stacey Pasco

Brent Deschamp was a doctoral candidate in civil engineering and mathematics when he entered the AAAS Mass Media Science & Engineering Fellows program. He wanted to broaden the public's perception about mathematics and thought the program could help.

Deschamp excelled at WOSU radio in Columbus, Ohio, where he was placed in the summer of 2005 to work as a science reporter. He quickly developed into a solid journalist whom the staff trusted to turn out multiple pieces with little hands-on management.

Journalism came easy to Deschamp, but his decision to complete his doctorate and return to teaching was also easy. That was his passion. As the manager of the program, I had a conversation with him in the last week of his fellowship; we discussed how he could carry his lessons about science journalism forward into his academic career. I assured him that they would be lasting. Deschamp wasn't so sure; nevertheless, he had enjoyed his summer as a reporter.

The year after his fellowship, Deschamp and his adviser were being interviewed by his university's alumni magazine about their research. It was apparent the reporter did not have a science background and was unclear on the types of questions to ask. Deschamp's adviser was excited that someone was taking an interest in their research, but was unprepared to answer questions effectively. Deschamp watched uncomfortably as his adviser tried to convey too much information. His answers were too long, used too much jargon, and simply overwhelmed the reporter. When it was Deschamp's turn to be interviewed, he tried to clarify some of the stronger points, but the reporter was already disengaged.

"The entire experience really illustrated why the AAAS program is needed," Deschamp later wrote to me in an e-mail. He'd learned what countless program Fellows before him had learned: Journalism may be the vehicle, but creating science communicators is the end goal of the program.

FELLOWSHIP EVOLUTION

The AAAS Mass Media Science & Engineering Fellows program seeks to increase public understanding of science and technology by increasing communication skills in young scientists. By exposing reporters and editors to the expertise of student scientists, while also training those same students in the nuances of journalism, the program can also improve the quality of information that is disseminated.

Each year AAAS places fifteen to twenty graduate and postgraduate science, engineering, and mathematics students at media sites nationwide to work as science reporters for the summer.

Past sites have included the *Los Angeles Times*, the *Chicago Tribune*, *Scientific American*, *Milwaukee Journal Sentinel*, National Public Radio, and *Voice of America*.

Some students want to immerse themselves in journalism to see if it is a career option, while others are looking to increase their communication skills to further the public's understanding of science.

The fellowship program has graduated more than five hundred alumni. They are part of a rich history of science-communication training that began in 1973 when the Russell Sage Foundation asked AAAS to take over management of its two-year-old social-science journalism summer internship. It quickly became apparent that the program should enlarge its scope to include the natural sciences. In 1982 AAAS made the decision to recruit more engineers to the program by adding "engineering" to the former "Mass Media Science Fellowship" program title.

Initially funded largely by the Russell Sage Foundation and a grant from the National Science Foundation, the program is now funded by AAAS and its affiliates and by corporations and foundations. When seeking funds from the National Science Foundation in 1980, AAAS staff made the case that "a more knowledgeable and well-informed public will permit more responsible decision-making about major issues of personal and public concern." At the time, the proposal cited media confusion after the 1979 accident at the Three Mile Island nuclear power reactor as evidence of the need for better media coverage of science.

Today issues such as stem-cell research, global climate change, and creationism in the science classroom occupy the front pages and public discourse. The public is often intrigued by "hot topic" research agendas, but it must understand the working parts of the entire issue. If scientists are committed to research, they must also be committed to communicating their findings.

Fellows quickly learn to let go of the intricate details of science in order to master the art of journalism.

"It is the journalist's job to find the essence of each story," said 2000 Fellow Lisa Durso, "which means distilling months or years of scientific work in to a single twenty-inch story."

Still, the Fellows enjoy the challenge of clarifying their logic and even allow, as 2004 Fellow Eric Tytell found, that "in a scientific paper, you can often get away with poorer logic because you can use so many more words." Journalism affords no place to hide.

In the early stages of the fellowship, one of the first frustrations can be the Fellows' struggle to connect with their readers. The students often have an innate curiosity that has drawn them to the sciences, and as media consumers they will seek out the smallest scientific discovery and find it interesting.

Then they meet their editors. Knowing that readers often need to be drawn in to a story, editors insist that Fellows find the one thing that matters most to the reader or the listener. The Fellows, like 2007 *St. Louis Post-Dispatch* Fellow Amy Maxmen, must not only discern why the story is of interest, but why it interests people in a specific community and how it will affect their daily lives.

"The busy editors at the *Post-Dispatch* taught me to be engaging, because I'd have about sixty seconds of their attention when I wanted to pitch a story," Maxmen said. "Explaining a science story reminded me more of making conversation at a cocktail party than lecturing biology in the classes I had taught. The topic had to appeal on a personal level, having something to do with where the person is from, what they like to eat, what their fears are, how they have fun, or how they earn a living. I've learned to get to the point—namely, how it impacts the editor, who cares how it impacts the readers."

Science News in the Digital Age

The media, newspapers in particular, are facing challenges brought on in part by our digital age. Hard-copy science sections may be disappearing, but science is still in the news. Longtime program mentor Ashley Dunn, science and medicine editor at the *Los Angeles Times*, says, "The main benefit of having a Fellow is it allows us (his staff) to tackle more stories in the summer months — a lot of stories that normally wouldn't be covered."

This is not to say that the fellowship doesn't recognize the changing media environment. The ever-updated 24/7 news cycle has affected how outlets distribute the news. According to the 2008 National Science Foundation report, *Survey of Public Attitudes Toward and Public Understanding of Science and Technology*, the Internet is the preferred source when people are seeking information about specific scientific issues; and it is second, behind television, as Americans' primary source of information about science and technology.

Fellows are now consistently called upon to "Webify" their original print or radio stories, sometimes adding sidebars that give readers more information than was available in their morning paper or on the radio during their drive to work. Fellows create original Web content, especially for media sites that publish on a weekly or monthly basis and that see the Web as a necessary tool for providing current news for their audience. Blogs are more and more common, and program Fellows are able to guest-blog for reporters, often helping to increase the number of posts if a full-time reporter's schedule doesn't allow for regular blogging.

All of this illustrates that as the demands of a newsroom change, so will the fellowship experience.

Next Steps

At the end of ten weeks, program Fellows usually know how they want the media to fit in their career paths. Of reporting alumni, a little fewer than 50 percent go into some form of science-communication field, but this doesn't tell the whole story. Some students enter the program hoping to transition to journalism and find they love practicing science more than reporting it. Others try to channel the momentum of the summer into a full-time job in journalism or freelance science writing. Still others always planned to go back to science and are now armed with a toolbox to help them wade through the ever-changing funding waters with Criterion 2 or other public-outreach requirements for grant seekers.

When some Fellows leave the program, they often think they should do one specific thing with their training. It's not important to AAAS what career path they follow; it's that they're dedicated to accurately communicating the science of their work. As program manager, I want them to take away certain key points. If they're going back to the bench, I hope they don't fear the press and will cultivate a relationship with their institution's public-information officer. If they're transitioning into science communication, I hope they stay dedicated not only to telling the stories of science but to making them personal.

Each year provides AAAS with a new opportunity to influence the career path of a group of young scientists and to positively affect how the public connects to discovery. AAAS is proud of the role it has played in developing the public's understanding of science and looks forward to working with the next generation of communicators.

For information about AAAS Mass Media Science & Engineering Fellowships, go to http://www.aaas.org/programs/education/MassMedia/.

Abby Vogel, PhD, is a communications officer in the Research News office at the Georgia Institute of Technology where she writes about Georgia Tech research discoveries and developments. Vogel also serves as a member of the IEEE-USA Communications Committee and as an editor for IEEE-USA *Today's Engineer*. While she was a graduate student at the University of Maryland conducting biomedical optics research at the National Institutes of Health, Vogel was awarded a fellowship with the AAAS Mass Media Science & Engineering Fellows program to work at the *Richmond Times-Dispatch* as a science reporter.

In this essay, Vogel describes how her experience as an AAAS Mass Media Fellow changed her career path from bioengineer to science writer.

5

Reflections of an Engineer/Science Writer

Abby Vogel

Being a science journalist is difficult. It's difficult to write without big scientific words, to make a science topic interesting to the average reader, and to ask the right questions during an interview and get an interesting quote.

I faced and conquered these challenges during the summer of 2005 as an AAAS Mass Media Science & Engineering Fellow, sponsored by IEEE-USA. The fellowship was a huge change from my "real" life as a doctoral student in biological resources engineering at the University of Maryland and a researcher at the National Institutes of Health.

The AAAS program offers student scientists and engineers a chance to get out of their labs and work as reporters at newspapers, radio stations, or television stations across the country. It aims to help them sharpen their ability to communicate complex scientific issues to nonscientists and help improve public understanding of science.

I applied for the fellowship because I love science and I love writing. And I knew early on that I might not be cut out for a job in academia because I was always more interested in learning about other people's research projects than digging more deeply into my own.

In my ten weeks as a science reporter at the *Richmond (va) Times-Dispatch*, I spent every day working on a new story and learning about new scientific research. During my fellowship, I published fourteen articles on topics as diverse as seismology, paleontology, physics, astronomy, health, birds, nail salons, wind farms, transportation, the space shuttle *Discovery*, and green-building techniques. The job was difficult but worth it when readers responded to one of my articles with phone calls and e-mails commending me on informing them about a topic that they were unaware of before and asking where they could find more information.

I received my first assignment on my first day at the paper. The next issue of the journal *Science* was going to include an article about earthquakes written by researchers from Virginia Tech. I had to interview the scientists and then find another expert in the field who could verify that the study was novel and brought new information to the public.

The next week NASA Langley held a media day for reporters to tell them about research at the Hampton, Virginia, facility—research that would hasten the space shuttle's return to space after the 2003 *Columbia* disaster. Seeking an angle that interested me, I talked to many researchers. And I found one using thermal cameras to see cracks and impact damage on the outside of the shuttle during a spacewalk. I even got to interview an astronaut to get his perspective.

That summer I also interviewed a Lynchburg College professor who used mathematical models to predict the winning times for each stage of the Tour de France and a William and Mary biologist who found that female birds were attracted to a mate with the same characteristics as the mates their female friends had chosen.

The most challenging assignment I received was to attend an Environmental Protection Agency workshop aimed at teaching nail technicians about the dangers of salon chemicals. The catch: The workshops were taught in Vietnamese. But I had to get the story, so I sat through hours of Vietnamese warnings about the hazards of working in a nail salon. My hard work paid off—the story landed on the front page of the newspaper the following day.

An interesting story arose when one of the senior editors asked me to find out when the Virginia Department of Transportation would repave the highway he drove on to get to work. Even though I didn't know why it was my job to find this out, I started calling around to people who worked at the Virginia DOT and found out there's actually a science to road repaving. I turned my editor's query into a fascinating and relevant story about how workers have to drive on all the state highways counting the cracks, ruts, and patches to determine which roads need to be repaved. I received the most responses from readers about this story.

My summer as a reporter let me keep science in my life every day without having to spend time in the laboratory to get results. The experience deepened my desire to be a science writer, but I couldn't just go out and become a full-time journalist because I had finished only my first year as a PhD student. But I knew I had to keep writing if I wanted to pursue science communication as a career after graduation.

With the articles from my fellowship in hand, I began writing freelance articles for a few newsletters at the NIH, where I was conducting my PhD research. I also took journalism classes at the University of Maryland and joined the IEEE-USA Communications Committee.

After the fellowship I wanted to learn about what science writers do at other types of media outlets, like universities. I learned that a university science writer writes press releases about professors' research that newspaper and magazine reporters read and may want to write about. Many universities also publish research magazines and annual reports that science writers contribute to.

During the summer of 2006 I started volunteering at George Washington University's Medical Center in the Office of Marketing and Communications. I loved it. The atmosphere at the university drew me in even more than the newspaper's. Instead of talking to researchers on the telephone, I was visiting them in their labs and offices and having one-on-one conversations about their research before it was published in a journal. Professors would call me to let me know they had a new journal article coming out, and others invited me to see some cool experiment going on in the lab. University communications was a much better fit for me. I had personal relationships with the faculty members and still felt close to the science.

In November of 2006, I started freelancing for the Georgia Institute of Technology's Research News and Publications Office. After finishing my PhD in August of 2007, I joined Georgia Tech as a full-time communications officer. In my current position, I write news releases about conference presentations, newly published papers, and faculty promotions. I also write articles for the Georgia Tech Research Institute's annual report and Georgia Tech's research magazine, *Research Horizons*. I currently serve as an editor for ieee-usa *Today's Engineer*, a monthly publication devoted to career and public-policy issues facing U.S. IEEE members.

Being a science journalist is difficult, but I learned it takes time, originality, and practice.

For information about former IEEE-USA–sponsored Mass Media Fellows, see http://ieeeusa.org/communications/massmedia .asp/.

Kristine Kelly is an account supervisor for Edelman Corporate and Public Affairs in New York where she helps hospital and pharmaceutical clients to publicize their research and discoveries. Kelly earned her doctorate in cell and molecular biology from Cornell University. After deciding that a career in research wasn't for her, she was chosen as an AAAS Mass Media Science & Engineering Fellow and intern at the *Richmond Times-Dispatch* in Richmond, Virginia. From there, Kelly went to Rockefeller University in New York as a science writer and public-relations officer. She joined Edelman in 2007. Apart from her interest in science communication, Kelly enjoys cooking, playing with her dog, and running. When she wrote this essay, she was also in training for the Chicago Marathon.

In this essay, Kristine Kelly, who considers herself a translator of science, provides a short course on how to explain complex information about science and technology to nonexperts.

Translating Science

From Academia to Mass Media to the Public

Kristine Kelly

When I introduce myself to a new client or colleague, one question inevitably follows: "How did a scientist end up working in corporate PR?"

It's a legitimate question: Going from earning a PhD in molecular biology to working in public relations is an unusual career move.

However, because now—more than ever—scientists need to be able to talk about their research and explain its significance, my career change makes sense. While I may not work at the lab bench anymore, my background in science is still essential to my job. My ability to talk about science—whether with other scientists, my PR colleagues, or members of the media—is my most valuable skill.

Because I straddle this divide between science and communication, I think of myself as a science translator.

I'm amazed that people are more comfortable in buying and selling stocks on the Internet than in talking about genetics. People may be more at ease with topics like finance over science because some media outlets have given them tools to understand financial questions. Those same media aren't always as good about providing tools for understanding science.

Yet the public's exposure to science is greater today than ever before. Drug advertisements have swamped TV and print (Europe may soon have a dedicated pharmaceutical-drug television channel), medical conditions can be researched online, and science-related issues such as funding for stem-cell research are appearing on voters' ballots.

I thought I could make a difference acting as an intermediary, the translator who has stood on both sides and understands both science and media.

Translating, I think, is an apt word to describe what most science communicators do. Similar to any specialty, science has its own language that can be difficult to navigate for those unfamiliar with it. Researchers who spend the majority of their days with other scientists can easily forget that a language exists in which words like ribosome, gene knockout, and polymerase chain reaction (PCR) are uncommon.

A science communicator, public relations officer, or press officer can help scientists translate their work into that everyday language.

Being reminded that their audience is probably not as fluent in science as they are, researchers who are accustomed to speaking only science can help bring the conversation back into everyday language.

In explaining a study's importance, I have found that non-scientists are more receptive to a story when scientists relate it in their first language, the one they learned before beginning a PhD program.

One trick I use is to recall my days as a first-year graduate student. I remember listening to seminars that I barely understood because I had not yet grasped all the scientific jargon.

Now, when faced with a difficult scientific concept, I ask myself, "How can I distill the science, translate the language, in such a way so that my younger self, sitting in the dark auditorium, would understand it?"

That technique can be helpful when deciding how much detail is necessary: Is it important to mention ribosomes? Perhaps I can describe a gene knockout as a mouse that scientists have bred so it's missing a specific gene.

When I talk with scientists, I try to play the student, reminding them that when they were students their level of understanding was much lower. I will repeat to scientists what they have told me, but I do it in lay terms (as in how PCR is like photocopying).

By reminding researchers how to frame their thinking in their first language, I come away with a better understanding of what the scientist is researching, not to mention better quotes and sound bites from the interview.

These interactions also help prepare scientists to speak to journalists, kind of "media training lite." If researchers can't "de-science" or speak to the official communicator in everyday language, they won't be able to do that with journalists either, increasing the risk of miscommunication.

I know how hard it can be to break down the language barrier. When I first decided to move out of the lab and into science communication, I freelanced for the public affairs department of New York Presbyterian Hospital.

Although I'd never written a press release, the public affairs staff were impressed with my scientific expertise and were desperate for someone who understood the science behind the research they wanted to highlight. (In press releases, scientific language often buries the news that would interest the public.)

For my first assignment—on the discovery of a potential new drug target for tuberculosis—I interviewed the scientist, read his paper, wrote my best version of a press release, and sent it off. My contact in the public affairs department sent it back unedited; it was so bad I needed to start from scratch.

Despite my attempts to write something resembling a press release, everything I wrote was wrong. I was trying to cram so much science into the release that the main message got lost. A scientist would have understood the points I was making, but those details just confused a nonscientist.

Finally, out of desperation, I broke the examples down, looking at the underlying patterns and style in which they were written. In effect, I used the scientific method to find out what made them work. The process helped me

> To isolate the big-picture idea: The discovery of a new protein essential to the survival of the tuberculosis bug could be a good drug candidate.
> To isolate the background needed to understand the big picture: What is the tuberculosis bug, why do we need new drug candidates, and what makes the protein essential?
> To pick out quotes from the scientist that supported and explained that big picture.

I tried not to get bogged down in all the scientific details that didn't address the main idea.

Finally, I had gotten it. I'd needed to understand that writing wasn't simply putting a bunch of words on a page. Just as in writing a scientific paper, there were a rhythm, a style, and a process behind writing a press release. But the task required me to use a different language.

When I started writing that press release, I didn't think it would be as hard as it was. During my career, I have met a number of scientists who I think feel similarly—that writing is something anyone can do. I've had press releases come back completely rewritten by researchers who believed they understood better than I how to write them.

But writing is as process-driven as science is. My experience highlights the effort and skill it takes to write a truly engaging and lay-friendly press release or article. That was a hard lesson to learn.

To help scientists who are interested in communicating to the general public, science communications officers could give those researchers a checklist or question-and-answer document. The document will help prepare the scientist for an interview, both with the science communicator and with reporters. The content would reflect the elements of a good press release, while providing scientists with a protocol to follow so they know what is expected. Questions such as:

What was your hypothesis?
Please sum up the main finding of your paper in one sentence.
What motivated you to explore this subject?
Does it relate to any human disease?
How will the results affect the public, either immediately or in the long term?

While these questions won't provide all the details needed for the press release or article, if a researcher has gone over them before the interview, the science communicator will have a starting point. Moreover, the scientist has had the chance to begin thinking about what his or her "big picture" finding is and how to describe it in language that the public will understand.

Scientists already have some idea how to do this; almost every paper they write has one paragraph where they put their research into the bigger context. Usually located in the discussion section of the paper, this is where they talk about how their research is going to help cure cancer, fight disease, or lend insight to a genetic condition. I call this paragraph the "Mom paragraph" because it is probably close to what they tell their mothers, or parents, about the work they do and why it is important.

In one of my scientific papers, the Mom paragraph is:

> In summary, the evidence for compensatory or redundant roles of ADAMs in heart development further establishes ADAMs as important factors in mouse heart development. In addition, evidence for a major role of ADAM17, but not ADAM19, in shedding of NRGs b1 and b2 should stimulate new interest in uncovering other potential mechanisms that might explain the mechanism underlying the heart defects in ADAM19 mice. This result also raises interesting questions about the role of ADAM17 in processing and activation of these and other NRG isoforms during development. These findings have implications for congenital heart disease, the most common form of congenital defects in humans.[1]

This paragraph uses a lot of scientific language because it was intended for other scientists to read. But the paragraph also identifies the main point: These proteins are important for heart development. It tells the communicator that this is not a breakthrough finding; instead, it supports existing data and may suggest a mechanism for why these proteins are important. Overall, the research may help scientists understand the cause of congenital heart disease, a major health problem affecting a large number of people. Already the communicator has a sense of what the paper's main idea is and how it fits into the big picture of general health.

Locating that paragraph in the discussion (it's usually one of the last two paragraphs of the paper) can be a huge advantage for a science communicator. The scientist has already distilled his or her research, giving the communicator guidance on what the researcher thinks is the main point. This paragraph gives the science communicator a reason to talk with the scientist or write a press release, and it can help in framing questions about the topic.

Scientists who realize the importance of communication skills and use them in daily conversations can devise persuasive arguments that win grants and shape the research environment. In addition, talking about their science in the media also publicizes their work to other scientists. A 1991 study in the *New England Journal of Medicine* (NEJM) found that a paper published in NEJM was three times more likely to be cited by other papers if it was also mentioned in the *New York Times*.[2]

Science communicators can help scientists develop their communication skills and show them that the need for communication is only gaining in importance.

Many scientists tell me they resist speaking up because they're concerned that journalists will misunderstand, get it wrong, and portray the research negatively. This problem involves language: a scientist who doesn't speak in everyday terms and a journalist who is unfamiliar with the intricacies of the science.

Science communicators in public relations can help. They are in a position to help researchers talk about their work and to make them feel comfortable and proud of what they're doing. For me, understanding the language of science isn't just about knowing what a knockout mouse is, but understanding a researcher's mentality.

I think my greatest advantage is having been on both sides of the line, in both the scientist's lab coat and the communicator's navy suit. But translating science doesn't require a PhD. For example, a PR writer might ask to spend a week in a lab with a researcher with whom he or she has a good relationship. Learning a bit more of the day-to-day ups and downs is a great way to get a feel for what research is like.

On the flip side, all graduate students in science could benefit from a communications, scientific writing, or other writing class in order to learn how to translate their science into language the public can understand. Not only will such a class teach students to understand and respect the writing process, it can help them understand the need to communicate and assist them in writing manuscripts and grants.

The greatest threat to science is miscommunication, which can lead to reports that frighten, anger, and mislead people. Such misunderstandings eventually cause a backlash onto the science community. But if researchers work with their press officers and the media to ease the language barrier and get it right, the results benefit everyone.

NOTES

1 K. Horiuchi et al., "Evaluation of the Contributions of ADAMS 9, 12, 15, 17, and 19 to Heart Development and Ectodomain Shedding of Neuregulins Beta1 and Beta2," *Developmental Biology* 283 (July 15, 2005): 459–71.
2 D. P. Phillips et al., "Importance of the Lay Press in the Transmission of Medical Knowledge to the Scientific Community," *New England Journal of Medicine* 325 (October 17, 1991): 1180–83.

Warren Leary reported on science for the *New York Times* for twenty years before retiring in 2008. During his last decade at the *Times*, he focused on covering NASA, including reporting on dozens of human and unmanned space flights and on space policy issues. Before joining the *Times*, he spent more than fifteen years covering science, technology, and medicine for the Associated Press. Leary has received state, regional, and national awards from the Associated Press, the New York State Press Association, and the National Association of Black Journalists. Leary has an undergraduate degree in journalism from the University of Nebraska–Lincoln and a master's in journalism from Columbia University.

Warren Leary has said he sees a need for more science education and reporting because the news is full of subjects such as global warming, stem-cell research, health concerns, and "miracle cures," making it difficult for people to sort out fact from fiction. In this essay, Leary argues that, despite the changes occurring in news media, journalism remains the best source of unbiased and accurate information about science; indeed, in Leary's view, beyond formal schooling, science journalists will continue to offer the best continuing education on science, no matter how new media shape the news.

7

Building New Media's Science Information on the Pillars of Journalism

Warren Leary

Science journalism—as with most journalism—is undergoing a crucial transition in response to technical changes affecting how the news and other information are presented. In addition, science journalists are adjusting to changes in the lifestyles, reading habits, and perceptions of their readers. Despite all these changes, the values of journalism can persist.

The basic role of journalism in our lives is to inform. An essential adjunct to that mission is to educate. In no area of this profession are these charges more linked than in the specialty of science journalism.

Many people find science and technology intimidating and difficult to grasp even as they influence almost every aspect of our lives. Science journalism—which covers medicine, health, social sciences, the environment, electronic and nanotechnology, physics, chemistry, robotics, engineering, earth and space sciences, life sciences and more—is an essential bridge to all of this information.

Sources of science information are both shriveling and expanding. Newspapers, magazines, television, and radio are providing less while Internet-based forms of communication are offering more. Dedicated science sections in newspapers and magazines continue to diminish as circulation numbers decline, while Web sites devoted to topics such as particular diseases, health-and-lifestyle trends, environmental issues, science policy, or science specialties increase in number.

When it comes to science news, the mass media serving the largest number of readers and viewers increasingly tend to focus on a few timely or sensational issues, such as global warming and potential pandemics, or "news-you-can-use" medical or health topics like dieting, fitness, disease prevention or therapy, and childhood illnesses.

Furthermore, many readers and viewers tend to want information more quickly and in a more compact form. Feeling the time pressures of modern life, fewer people have either the time or patience to read a daily newspaper or to adjust their routines to accommodate scheduled newscasts.

Results of these trends are not necessarily all negative. When focusing on a popular issue, such as unsafe food products or drugs, the news media draw the public's attention for a time and can influence reforms that meaningfully address a problem. However, issues that require sustained attention, such as long-term environmental cleanups, may suffer when public focus shifts to something else.

Increasingly, news is becoming the purview of "aggregators" who collect and collate information from many sources and present it in one place for those interested in a topic. People may now "customize" their news interests by creating Web pages that display only news on preselected topics — a practice dedicated to the conceit that most people know in advance what they need to know and don't have the time or inclination to deal with unexpected or random information.

Sources of science and technology information on the Web include pages and sites from established news entities such as newspapers like the *New York Times*, magazines like *New Scientist*, *Science*, and *Discover*, and broadcast outlets such as National Public Radio. In addition, colleges, universities, and major hospitals post science and medical information on their sites, as do non-profit health-and-disease organizations and some professional medical specialties.

In this Internet mix, blogs now address almost every topic, including many hosted by scientists, physicians, and other specialists who are seizing this opportunity to explain their work to the public.

With the proliferation of science-news sources and trends toward do-it-yourself newsgathering and distribution, science journalists can and should play a role. They can apply traditional journalistic values to the new media and convey to the public why those values are important and necessary. This contribution should include reinforcing the journalistic pillars of perspective and balance.

A major challenge to consumers is how to make sense of so many sources of information. How does one prevent this growing mound of information from becoming a proverbial Tower of Babel?

Information can represent both a tool and a weapon, a path to truth or to deceit. Feeling ignorant or uncertain about a topic and wanting correct answers to questions, people seek data and facts.

Few go looking for misinformation and uncertainty—which, unfortunately, are all too abundant in the oceans of information in which we are now submerged.

People now roam the Web with abandon looking for information. Guided by search engines or referrals from friends, they go from site to site looking for everything from physician recommendations to details about the dangers of radon gas in their homes.

But can they trust what they read on their computer monitors or what they hear on podcasts?

Here lies the role of the science reporter and of good journalism, in general. Part guide, part referee, part adviser, part educator, a science reporter can play a major part in making sense of it all. Science reporters act as objective, honest brokers in analyzing information about science and technology and presenting it to the public—even a reluctant public that thinks it isn't interested in science and other technical things.

For years, professional journalists have discussed and debated the concept of objectivity in news-gathering. Can any intelligent person remain truly unbiased, and therefore objective, after sorting through the facts and covering a topic in detail? Can he or she form an opinion on an arguable topic without conveying the opinion while reporting and writing on that topic?

A traditional tension in newsrooms and journalism classrooms has involved the goal of maintaining objectivity. Editors and teachers demand that reporters and students do more reporting when stories seem too one-sided or not enough effort has gone into gathering opposing views or opinions. Reporters and journalism students rewrite stories to provide more balance or differing perspectives, knowing that objectivity is a worthwhile professional goal in the best interest of readers and viewers.

The issue is not about whether journalists have opinions;

instead, it is about whether they try to keep an open mind and actively seek to find and report a variety of viewpoints. The point of this effort is to present readers with honest representations of differing opinions.

Too often, those writing in alternative media haven't prioritized either objectivity or accuracy. Bloggers commonly editorialize and give short shrift to opposing views. Some Web sites provide what appears to be straightforward information without identifying the sponsors or their intent.

The question isn't whether traditional print and broadcast media are "good" or the best way to present information, and alternative forms such as blogs, streaming videos, or podcasts are "bad." Traditional media have gained from using new tools, such as innovative graphics, to enliven their Internet presentations and enhance their core products. Whatever the means of conveying information—whether about science or any topic—the information must honestly serve the reader or viewer and use the medium's strengths to achieve that goal.

Much information distributed by new media is the product of small groups of people or solo practitioners. If they're serious about striving for accuracy and balance, they might consider maintaining or adopting some traditional journalistic practices, such as using copy editors. Aside from corrections in grammar and punctuation (which most readers will appreciate), articles and other presentations will improve by having someone else look at them before publication or posting, asking questions and sometimes challenging assumptions.

Done properly, copyediting doesn't have to be a time-consuming process that cripples the immediacy of putting things out on the Internet. In fact, the best way to protect accuracy in the media, old or new, is to get it right the first time. Advocates of new media often suggest that the Internet is self-correcting, noting that readers can easily post comments and challenge misinformation. However, bad information—much like gossip—tends to perpetuate on the Web, picked up and spread by passersby who may not notice or care about comments that challenge or correct.

Politics, self-interest, conflicts of interest, prejudice, or extreme dedication to an issue can result in the distortion or misrepresentation of knowledge about science and technology.

Controversial issues such as evolution, environmental cleanup, global warming, embryonic stem-cell research, and mass immunizations to thwart epidemics point to the need for the public to have clear and unbiased information about science in order to make informed decisions.

These issues won't go away, and new ones arise with increasing frequency. Decisions concerning science, made either by voters or by their representatives, touch the lives of all people, even if they aren't paying attention.

Some may feel they can ignore science and technology. But when these subjects seize their attention, their level of science education—much of it acquired through the media—may determine the impact on their lives. For example, when a company develops an antiviral vaccine that may prevent future cases of cervical cancer, the public may not pay much attention. But if a school system mandates that all young girls be vaccinated, even though there are no studies of long-term risks, parents with daughters suddenly get interested and want answers.

Science education isn't just for specialists or nerds. It is a key to an informed, healthy, and safe life. Journalism has a huge role to play in science education for people who never took science courses or didn't pay much attention if they did. While science literacy or education might not seem important in everyday life, it can, in fact, be a matter of life and death.

Each day, people die because they don't know or don't appreciate the role of scientific principles in their lives. A man cleaning the gutters of his house falls when he reaches too far from the top of a ladder. A speeding driver thinks that, because of his skill behind the wheel, he can take a curve posted for forty miles per hour at seventy. A woman trying to clean a stubborn stain in a confined bathroom succumbs after mixing ammonia and chlorine bleach.

Outside of formal schooling, journalism is the primary source of continuing science education for many. A magazine article, a television news spot, or an Internet blog can be an alternative science textbook, providing critical information. Someone—journalists traditionally—should strive to make it the best and most accurate possible.

Science journalists help sort out, interpret, and distill voluminous amounts of information from dozens of sources and present it in a way that gives the public some confidence that it has the best information available.

This role has been the past and is the future for journalists who write about science.

Boyce Rensberger has been a science writer or science editor for more than thirty-two years, beginning at the *Detroit Free Press*. From there he went to the *New York Times*. Then he freelanced and became head writer of a PBS science series for children, *3-2-1-Contact!* In 1981 he became senior editor of *Science 81–Science 84*, a popular monthly magazine published by the American Association for the Advancement of Science. In 1984 Rensberger went to the *Washington Post*, where he served as science writer and science editor. He became director of the Knight Science Journalism Fellowships at MIT in 1998 and retired from that position in 2008.

Rensberger's books include *Life Itself: Exploring the Realm of the Living Cell*; *Instant Biology: From Single Cells to Human Beings, and Beyond*; *The Cult of the Wild*; and *How the World Works: A Guide to Science's Greatest Discoveries*.

In this essay, Boyce Rensberger draws on his long experience as both a journalist and a teacher to explain how scientists should respond to journalists who want to write about the scientists' work. But science journalists are becoming endangered species in newsrooms, meaning science gets minimum coverage at a time when citizens' understanding of science couldn't be more critical. So Rensberger's advice is for the lucky few scientists who will actually have opportunities to talk with a reporter about their work.

Preparing Scientists to Deal with Reporters

Boyce Rensberger

Scientists and science students need to prepare themselves to deal with the news media.

Like it or not, lay-language newspapers, magazines, and television programs form the only significant media through which adults in the general public learn of current research and about possible controversies related to it. These media help shape the public discourse on science, which, in turn, influences the decisions of government agencies that may fund or restrict research.

Moreover, it is increasingly accepted that scientists who pursue their work using tax dollars have an obligation to report to their ultimate funders what they are doing with the money.

One step in preparing for this effort is to form a more accurate impression of what the American public thinks about science.

Good evidence on this topic has come from the National Science Foundation's biennial surveys of public attitudes toward science and public understanding of scientific concepts and facts.

Perhaps the finding most counterintuitive to many scientists is that the public overwhelmingly admires scientists and believes that the vast majority of scientists want to improve life for ordinary people. Tallying a variety of surveys on the point, the NSF found that between 85 percent and 90 percent of adults agree that "developments in science helped make society better."

Moreover, most people believe scientists *want* to make the world a better place. Some 89 percent agree that "most scientists want to work on things that will make life better for the average person." No "mad scientist" stereotypes here.

Other responses to the NSF surveys support the conclusion that the public wants more information about science. Some 58 percent say their interest in science is high, but only 10 percent say they feel well-informed about science. In other words, the public is telling survey researchers of its interest in science, but it isn't getting enough information to satisfy that interest.

Scientists can help fill this need, but most don't have the time or skill to convey their research to the public directly. So we science journalists can be conduits. But for that transfer of information to happen, scientists need to appreciate the constraints under which journalists work.

Typical science journalists cannot specialize in any one sub-specialty of science enough to become totally conversant with all of it. A newspaper science reporter may write about astrophysics today and molecular biology tomorrow and be plugging away on a story on paleontology for next week.

We know the risk of making factual mistakes is high. Although we are loath to make errors, we want to hear from you scientists if you spot one. Don't hold back, unless your complaint is that our story didn't say everything there was to say about your subject.

We work under severe constraints on both time and space. We often have only a few hours or a few days to do a story, and three typewritten pages, double-spaced, is the length typically allotted to a newspaper story.

So help us communicate your science to the public. If you have a paper coming out in one of the more visible journals, prepare yourself for a phone call from a reporter. Here are a few steps:

1. Be ready to explain why anybody outside your field should care about the results of your work. Most reporters are looking for scientific developments that are either important (they will have a practical impact on the lives or fortunes of ordinary people) or fascinating (they fill us with wonder or amazement about the natural world). Remember that the public feels no obligation to read about your work; you need to help the reporter give people a reason to read the story.

2. Think of ways to explain the technicalities of your research in everyday language. Quantum physicists have a hard time following advances in molecular biology, and vice versa, unless the explanation is put into plain English. Imagine how much trickier such stories are for mere lawyers, truck drivers, or historians to follow.

One way to rehearse your translation is to imagine how you would explain your finding to your spouse or parent. In fact, you might even do that experiment.

3. Be prepared to explain how your work fits in the overall progress of your field. Are you the first to find a result like yours? Are you extending or confirming the work of others? Did some key question in your field remain unanswered until you made your finding? Does your work nail it for all time, or will confirmation be necessary from another lab or another line of study?

We journalists know that science almost always involves a step-by-step process over many years. So we don't want to look like fools by portraying your finding as having emerged full-blown from the vacuum. Nor do you want to be made to look foolish.

4. Be ready to discuss uncertainties in your findings. Good science journalists understand and respect error bars.

5. During the interview, feel free to ask the reporter whether he or she understands what you're saying or would like to have you say it another way. Reporters, especially young ones, are reluctant to look like idiots in front of scientists, so they may pretend to follow what you say. You don't want them to hang up or leave your lab puzzled about something.

But if the interview starts with a vague question like, "So what's your research about?" be afraid. You have encountered a reporter who hasn't done the minimum amount of homework before making the first call. You may feel free to tell the reporter to do more background reading first, unless you have a lot of time or the reporter is new at your local paper and you'd really like him or her to cover your work—someday if not now.

If you can refer the person to some fairly simple background reading, that would be great. Even better, if you want to cultivate this reporter, invite him or her to your lab for a brief tour or to have lunch with you or your colleagues. This kind of personal interaction—away from deadlines—can build mutual trust and vastly improve your chances of the resulting stories being ones that really do satisfy the public's wish to learn more science.

David Ehrenstein edits *Physical Review Focus* (focus.aps.org), a Web publication of the American Physical Society. The site explains in plain English selected recent physics research for a broad audience of students, journalists, and physicists. Ehrenstein has a PhD in physics; following his postdoctoral research at NIH, he worked for six months as an intern journalist at *Science* magazine. His shift to science writing began during graduate school when he realized his strongest skills and interests were in interviewing a wide variety of scientists and explaining science. Rather than specializing, he also liked learning about many areas of science.

In this essay, David Ehrenstein offers several examples of how he and his writers have used mental imagery—analogies, really—to explain complex physics concepts in stories published in *Physical Review Focus*. Speaking at the 2007 UNL conference, he said his "mantra" for *Focus* has been "A mental image is worth a thousand equations." Ehrenstein, however, anticipated objections that such techniques oversimplify the physics and don't tell the whole story. In response to that complaint, he remarked, "Perfectly accurate is perfectly inaccessible," noting that the details of the mathematics connected with physics are likely to drive away nonexpert readers.

Picture Power

David Ehrenstein

Even the most abstract, theoretical physics research is ultimately about something physical in the real world, not just mathematics. So I've found that if you look hard enough, you can almost always find a mental image to replace a lot of the physics language and equations — just don't expect the physicist to come up with it.

The image usually takes fewer words to explain than the abstract concepts and provides much deeper understanding for readers.

One of our stories described a newly explained phenomenon where two floating sheets of sea ice collide and join with an interleaved pattern of "fingers" (focus.aps.org/story/v19/st6). The physicists said the phenomenon was caused by an "instability" involving "buoyancy" and "elasticity."

Rather than explain that mouthful, we used the image of a

very long air mattress floating in a pool. To hoist yourself onto it from one of the long sides, you have to pull the mattress down in one spot, which then raises the sections to your right and left. If two floating ice sheets meet with some force, and a projecting bump from one manages to push on top of the other in one spot, both sheets end up with the air-mattress-like wave along their edges. But one sheet is up where the other is down, so they can cut "slots" into each other to make an interlaced pattern. The air mattress image conveys a lot of physics that would be hard to explain otherwise.

The truth is, nonscientists already know a lot of physics, but our brains work a lot better with pictures than with words. So the image allows the text to skip over wordy, textbook definitions for terms like "instabilities" and "elasticity." Mental images also help make a physics story come alive as the reader imagines the scenario—air mattresses in swimming pools are a lot more exciting than the floating sheets of wax the scientists used to study the phenomenon.

Another story described a technique for following the motion of tiny, oil-filled particles in water flowing through a half-millimeter-wide tube (focus.aps.org/story/v20/st21). To some extent, the particles were expected to mimic red blood cells flowing through blood vessels.

The researchers were trying to determine whether the particles simply "went with the flow" and remained on parallel tracks—or did they also move somewhat perpendicular to the flow? To find out, the researchers used a technique called pulse gradient spin echo, which involves applying brief pulses of magnetic fields and then interpreting the electromagnetic waves emitted by the atomic nuclei in response.

In this case, too, we came up with a simple image: cars on a highway, where each lane has a different speed limit. The image contained the essential ideas but was quite different from the physicists' explanation. We explained that the researchers "found a surprise: even in slowly flowing, dilute colloids, the particles refuse to remain on purely parallel paths, like calm drivers on a highway. Instead, they change lanes erratically."

The series of four magnetic pulses applied by the researchers was the equivalent of starting six cars at a starting line, allowing them to drive at different speeds for a fixed time, and then forcing all of them to reverse direction for the same time period. If the cars didn't end up synchronized at the starting line, then at least one wasn't following its speed limit.

The research team was surprised to find that for measurements where the whole cycle was longer than about one hundred milliseconds, the "cars" were out of synch at the end. This meant that particles had strayed from their "lanes" and their associated speeds—fastest at the center and slowest near the walls. The result was unexpected because dilute particles in flowing water were thought to remain entirely within the fluid equivalent of lanes.

In reality, the particles never physically moved backwards; the flow continued unchanged during the entire experiment. The technique uses magnetic field tricks somewhat similar to those used in medical MRIs to track the spinning motion of the atomic nuclei in response to those fields. But explaining the nuclear and magnetic physics would have taken us far afield for relatively little gain.

In both of these examples, we needed a fairly deep understanding of the physics before we could come up with good analogies. It can take some persistent questioning of the scientists to reach this level of understanding, but it's worth the effort.

Often, journalists covering a physics story don't attempt to get a deep understanding or to convey the ideas to readers, preferring to stick with the motivations and conclusions of the research. In some cases, journalists are put off by theory and equations.

Failing to understand the math, journalists may assume that it doesn't contain any information worthy of readers' attention.

One story that might have been in this category was about wrinkling (focus.aps.org/story/v11/st7). The physicists came up with a theory to predict the wavelength and size of wrinkles in certain situations; the theory could potentially be applied to manufacturing of smooth materials in many industries. Rather than simply explain the question (what's the wavelength?) and the results (comparison between predictions and experiments), we wanted to explain the ideas behind the equations.

The equations expressed the competition between two "interests" whenever a thin sheet is supported by a stiff foundation. In the case of a shrunken, dried-up apple with a wrinkled skin, the apple's flesh is relatively stiff and prefers the smallest waves possible. On the other hand, the skin, like any thin sheet, prefers one large wave to many small ripples — just imagine trying to keep a ribbon compressed into many tight S-turns rather than a single loop.

Starting with a mathematical expression for these two competing tendencies or interests, the theorists correctly predicted the size and wavelengths of wrinkles on an apple and the back of a human hand.

This kind of intuitive explanation using imagery helps to demystify the whole business of theoretical physics; the math is just a representation of real-world behavior.

Another common problem in physics stories is the concept of energy. Physicists are always explaining phenomena by saying that a system tries to lower its energy. I prefer to illustrate this rather abstract principle with the more intuitive notion that objects tend to roll downhill, which is a familiar example of energy-lowering.

One example was a story on a new technique for slowing and ultimately stopping atoms in a beam (focus.aps.org/story/v21/

stg). The beam was directed through the centers of sixty-four coils of wire, each of which was an electromagnet that could be quickly switched on and off as a group of atoms came through. We described the magnets as acting like hills the atoms had to climb; the climbing slowed them down. Each "hill" was quickly switched off just as the atoms crested it, so the atoms would not speed up again by sliding downhill.

Mental images can even help explain physics principles that are unfamiliar, such as quantum mechanics. In our story on the 2007 physics Nobel Prize (focus.aps.org/story/v20/st13), we described electrons that can be only spin-up or spin-down. In our explanation, they scatter from atoms in a magnetic layer only if the magnetic field in that layer points in the same direction as the electron's spin. This purely quantum-mechanical effect is not at all intuitive, but it leads directly to the principle of giant magnetoresistance—the basis for all modern hard drives—for which the prize was awarded.

To explain the concept, we drew a mental image of a sandwich of these magnetic layers with electrons zipping across in all directions. Readers will go along with the explanation as long as you tell them exactly what picture they should have in their heads, no matter how weird.

Mental images are a great way to make complicated concepts seem simple, skip over wordy explanations, make stories come alive, and allow stories to jump straight to the physical principle without a lot of preliminary terminology. If a physics story seems wordy and abstract, it can usually be improved by adding a clearly drawn mental image.

Sidney Perkowitz earned his doctorate at the University of Pennsylvania. As Candler Professor of Physics at Emory University in Atlanta, Georgia, he has produced more than one hundred research papers and books in areas of condensed matter physics, biophysics, and optics, including semiconductors, superconductors, biomolecules, lasers, and submillimeter spectroscopy. His popular books *Empire of Light*, *Universal Foam*, *Digital People*, and *Hollywood Science* have been translated into six languages. He writes for various periodicals, and his media appearances include CNN, NPR, and the BBC. His book in progress is *Monochrome: Adventures in Black and White*.

   Eddy Von Mueller, PhD, is a lecturer in film studies at Emory University, where he teaches courses in film and television history, Asian cinemas, and narrative filmmaking. He is widely published in the popular press as a media critic and commentator and has written scholarly articles on the American film industry in the first decades of the twentieth century and on the work of acclaimed director Akira Kurosawa.

In 2006 Sidney Perkowitz and Eddy Von Mueller introduced the course "Science in Film" at Emory University. In this essay, they explain that science-fiction films do, in fact, contain some accurate science and that viewers can, to some extent, learn about science through such films. In keeping with the focus of this book, this essay explores one way to engage nonscientists — in this case, students and the general public — to pay attention to science.

10

Communicating Real Science through Hollywood Science

Sidney Perkowitz and Eddy Von Mueller

If you look for science in Hollywood films, you'll find it mostly in science fiction. But although science fiction films are plentiful and viewed by millions, they are rarely considered as channels for communicating meaningful science.

Even the wildest science fiction film, however, has somewhere in it a nugget of real science. The science may be sadly distorted, but as our course Science in Film and other efforts show, Hollywood's science fiction films can still be a remarkable resource to convey both science and a healthy critical perspective to students and the general public.

Since Georges Méliès's pioneering 1902 effort *Le Voyage dans la Lune* (*Voyage to the Moon*, based on Jules Verne's story about a cannon that shoots a space capsule to the moon), more than 1,400 science fiction feature films covering most areas of science and technology have been released in the United States.[1]

These films have become a cultural force, ranking among the most widely seen and most profitable of movies. Science fiction works such as 2001: A Space Odyssey (1968) and the original Star Wars (1977) appear on lists of all-time best films.

They inspire young people to become scientists. From "wormholes" to "Beam me up, Scotty," the ideas and vocabulary of science fiction films have entered our thought and language. And these films reflect contemporary issues of science and society—from Cold War anxieties about nuclear disaster to today's concerns about genetic engineering.

But how accurately do these films present science? James Cameron, who wrote and/or directed Aliens (1986) and the Terminator series (1984, 1991, 2003), has said that Hollywood science fiction movies "almost never get their facts right . . . they always show scientists as idiosyncratic nerds or actively the villains."[2]

Essayist Susan Sontag has made a similar point, but adds that science fiction films "can supply something the [science fiction] novels can never provide—sensuous elaboration . . . by means of images and sounds."[3] That powerful "sensuous elaboration" as enhanced by elaborate special effects makes science fiction films compelling, but also encourages filmmakers to hype the science for dramatic effect.

Dramatic appeal, however, makes science fiction films effective additions to regular science courses; for instance, Physics in Film, taught by Costas Efthimiou and Ralph Llewellyn at the University of Central Florida.

One film used in this course is Armageddon (1998), in which a huge asteroid "the size of Texas" streaks in on a collision course toward Earth (fig. 1). NASA sends up a crew of oil drillers in a spacecraft to implant a hydrogen bomb in the object. The explosion splits the asteroid into two pieces that fly off on diverging paths and miss Earth.

FIG. 1. *Armageddon* (1998). Science fiction films offer opportunities to teach science by analyzing what's shown on screen. In *Armageddon*, for instance, students can calculate the validity of a proposed scheme to save Earth from an incoming asteroid, or the probability of a space rock hitting a particular spot on Earth (here, Manhattan's Chrysler Building). Touchstone/ The Kobal Collection.

But as Efthimiou and Llewellyn write, when the class quantitatively analyzes the situation according to the laws of mechanics, "The students are astonished. Instead of being hit by one Texas-size asteroid, Earth will be hit by two half-Texas-size asteroids about 400 meters apart!"

That is, even a hydrogen bomb has insufficient power to push the two huge chunks of matter very far apart. Efthimiou and Llewellyn report that their combination of film drama and rigorous analysis keeps student interest high, and that students in the film-oriented course perform better than those in a nearly identical course without films.[4]

Our cotaught course at Emory, Science in Film, is different. Rather than working within an established scientific syllabus, it combines scientific and cinematic analyses. The textbooks are Sidney Perkowitz's *Hollywood Science*, which emphasizes the science in science fiction;[5] and

TABLE 1. Semester topics and filmography for the course Science in Film

Week	Topic	Required films	Suggested films (partial list)
1	Historical Introduction	*Le Voyage dans la Lune*	*La Hotel Electrico; Destination Moon*
2, 3	Alien Encounters	*War of the Worlds* (2005); *The Day the Earth Stood Still*	*The Thing from Another World; The War of the Worlds* (1953); *Invasion of the Body Snatchers* (1956, 1978); *E.T.; Close Encounters of the Third Kind*
4	When Worlds Collide	*Armageddon*	*Deep Impact; When Worlds Collide*
5	Worlds Gone Mad	*The Day After Tomorrow*	*Soylent Green; Volcano; Waterworld*
6, 7	Smashing Atoms	*On the Beach* (1959); *Fat Man and Little Boy*	*Godzilla; The Sum of All Fears; The China Syndrome; Chain Reaction*
8	The Third Horseman	*Outbreak*	*The Omega Man; The Andromeda Strain; Panic in the Streets*
9	Send in the Clones	*Gattaca*	*Jurassic Park; The Sixth Day; The Boys from Brazil; The Island of Dr. Moreau* (1996)
10, 11	Men and Machines	*Colossus: The Forbin Project; Terminator*	*A.I.; 2001: A Space Odyssey; Westworld; RoboCop; I, Robot*
12, 13	The Movie Scientist	*Contact; Dr. Strangelove*	*Metropolis; Frankenstein; The Boys from Brazil*
14	Reel Science	*Kinsey*	*Gorillas in the Mist; Dr. Ehrlich's Magic Bullet; Infinity; A Beautiful Mind*

Vivian Sobchack's *Screening Space*, which focuses on the genre's cinematic and cultural meanings.[6]

The students examine a suite of films covering topics from climate change to alien life (table 1). Each topic begins with a lecture about the relevant science. For instance, "Smashing Atoms" covers the 1938 discovery of nuclear fission and the enormous energy it can release; the Manhattan Project, Los Alamos, J. Robert Oppenheimer, and Hiroshima; the effects of nuclear radiation; and fusion power. Armed with appropriate background, the students watch required and suggested films, then discuss and write about what they've seen.

Without emphasizing detailed calculations as in a traditional science course, our course helps students (both science and humanities majors) gain an appreciation of magnitudes like the distances between stars and of important scientific ideas, such as plate tectonics, global warming, and how a DNA molecule conveys genetic information.

Our humanities-sciences double-teaming encourages critical thinking because science fiction films package more than science; they also sell attitudes toward science and place it within a cultural, political, and textual framework. We teach the students to critically evaluate both the science underlying a film and how the film spins its science. Students see that science fiction cinema is part of a larger societal discourse that includes scholarly journals, textbooks, journalistic treatments, and public-policy debates.

Science fiction films can also be used to teach science to the general public, for whom these films provide a loud, if not always credible, voice. Hollywood's marketing machine can reach a far vaster global audience than any news organ, let alone any educational institution.

The interest piqued or anxieties stirred by hyperbolic science fiction can create a space for public scholarship about critical

FIG. 2. *The Day After Tomorrow* (2004). In this film that won awards for its intense special effects, global warming is presented as causing tsunamis (here shown inundating New York's 42nd Street Library), incredible storms, and an ice age. The film exaggerates the scale and pace of these phenomena, but the science behind global warming is real and points to a significant issue. 20th Century Fox/The Kobal Collection.

events like the launch of Sputnik in 1957, the publication of Rachel Carson's *Silent Spring* in 1962, and the cloning of Dolly the sheep in 1996.

As laws are passed and policies changed — from banning DDT to limiting federal funding for genetic engineering — films reflecting these realities are hustled into production and onto screens. Examples include clone-sploitation pictures like *Multiplicity* (1996), *The Sixth Day* (2000), and *The Island* (2005).

Science fiction films can significantly influence public perceptions, as illustrated by *The Day After Tomorrow* (2004). Starring Dennis Quaid as a heroic climate scientist, the film presents the supposed effects of global warming — extremely violent weather, a tsunami that floods New York City, and an ice age — via spectacular, award-winning special effects (fig. 2).

Though these hypothetical outcomes — including the ice age — aren't utterly incorrect, they are

exaggerated or shown as occurring far more rapidly than scientists actually predict.

In contrast, Al Gore's sober documentary *An Inconvenient Truth* (2006) closely follows the reputable science of climate change (fig. 3). As documentaries go, this film has been a huge success, earning an Academy Award and a near-record box office gross of $49 million.

But *The Day After Tomorrow* had a box office gross of $543 million, reaching roughly ten times the audience.[7] Moreover, a recent study of viewers' reactions to the film shows that it had already significantly heightened public concerns about global warming well before Gore's film.[8]

This comparison highlights the immense power of science fiction cinema to educate people about science, but also reminds us that Hollywood values dramatic and emotional power over scientific accuracy.

FIG. 3. *An Inconvenient Truth* (2006 — Academy Award, Best Documentary Feature). Unlike *The Day After Tomorrow*, in *An Inconvenient Truth* former Vice President Al Gore presents the scientific evidence for global warming through images, graphs, and numbers. The combination of fictional and factual approaches can be an effective way to inform students and the public about science and its societal impact. Lawrence Bender Prods./ The Kobal Collection/ Lee, Eric.

Still, the dramatic can be balanced with the factual, both in classrooms and on bigger scales. For example, the TV show Num3ers reaches large audiences through its blend of real mathematics with engaging characters and stories. Organizations like the National Science Foundation, the National Academy of Sciences, and the Sloan Foundation are examining how best to use science fiction film to enhance scientific literacy.

The day may yet come when, settling back in your theater seat, you'll take in thrills, drama, and accurate science along with your popcorn.

NOTES

1 Data from IMDB, the Internet Movie Database, http://www.imdb.com/.
2 Andrew C. Revkin, "Filmmaker Employs the Arts to Promote the Sciences: A Conversation with James Cameron," *New York Times*, February 1, 2005.
3 Susan Sontag, *The Imagination of Disaster* (London: Vintage, 1994).
4 The quote comes from Costas J. Efthimiou and Ralph A. Llewellyn, "Cinema as a Tool for Science Literacy," http://arxiv.org/PS_cache/physics/pdf/0404/0404078v1.pdf/ (April 16, 2004). For more about the course Physics in Film, see Efthimiou and Llewellyn, "Avatars of Hollywood in Physical Science," *Physics Teacher* 44 (January 2006): 28–33.
5 Sidney Perkowitz, *Hollywood Science: Movies, Science, and the End of the World* (New York: Columbia University Press, 2007).
6 Vivian Carol Sobchack, *Screening Space: The American Science Fiction Film*, 2nd ed. (New Brunswick NJ: Rutgers University Press, 1997).
7 Box office figures from Boxoffice Mojo, http://www.boxofficemojo.com/.
8 Anthony Leiserowitz, "Before and After The Day After Tomorrow," *Environment* 46 (2004): 2–37.

John Janovy Jr., PhD, is the Varner Professor of Biological Sciences at the University of Nebraska–Lincoln, where he has taught for more than forty years. His research specialties are in protozoology, parasitology, and parasite ecology. He earned his PhD from the University of Oklahoma. An accomplished writer, Janovy has written many books, including *On Becoming a Biologist*, *Ten-Minute Ecologist: Twenty Answered Questions for Busy People Facing Environmental Issues*, *Keith County Journal*, and *Yellowlegs*. Janovy has said that he writes "to communicate the biological, scientific information in such a way that the average citizen can understand and appreciate the complexity of Earth."

Dr. Janovy responds here to essays submitted by other authors. In the process, he makes the case for the public knowing more about not only the outcomes of science but about its purposes and methods.

Afterword

The Challenge and the Need to Talk
and Write about Science

John Janovy Jr.

The authors of this book have addressed, albeit indirectly,
two of the most perplexing questions in American culture:
(1) What is science? and (2) Why should I be interested in any-
thing "scientific"?

These questions are perplexing because "science" can be
defined easily at the *Webster's Dictionary* level; but in practice,
science can hardly be defined at all. For example, astrophysics
is without doubt a sophisticated scientific enterprise, but then
so is the study of malaria immunology. The two activities have
virtually nothing in common except use of natural phenomena as
a subject of interest and the requirement for testable assertions
in their narratives.

The second of these opening questions also is easy to answer,
but again, the general phrase "because it's of personal economic

importance to you and your family" hides a staggering diversity of actual reasons, ranging from "for some stupid reason my kid loves it" to "I've just been diagnosed with cancer, and those researchers need to find a cure."

Finally, whether the authors intended to or not, they also have addressed what has become one of the most intriguing questions arising out of the culture wars currently raging in America, especially those involving evolution: What is the fundamental nature of scientific inquiry?

Abby Vogel's (p. 46) experience of writing "on topics as diverse as seismology, paleontology, physics, astronomy, health, birds, nail salons, wind farms, transportation, the space shuttle *Discovery*, and green-building techniques" typifies the challenge of a so-called science writer, that is, one who unabashedly tries to explain complex phenomena to the general public and therefore, by default, defines the term "science." Vogel's story on repaving Virginia highways also, however, contains a hidden narrative known well by legions of scientists whose research has no obvious practical application. We can hear some guy in a bar asking a Virginia Department of Transportation (DOT) worker what he does in his job, then picture this person's reaction at being told, "I drive around all day in a government vehicle and count cracks in the highways."

Suddenly eyeballs become tax-dollar signs; we hope a satisfactory explanation occurs before a fight ensues.

Every scientist knows that data collection is serious business and that the methods of doing so are defined by the problem to be addressed. This seemingly esoteric principle dictates our transportation worker's activity, but we hope that in this particular situation he has acquired a journalist's skill with words to complete the conversation—that is, to make the connections among:

Observation ("There are cracks in the concrete"),

A question derived from that observation ("How many cracks per mile make this highway unsafe?"),

Testable assertion ("This road needs to be fixed or the accident rate will increase"), and

Application ("Spend tax money; send out the paving crew; reduce the number of accidents")

Although the subject is as practical and mundane as highway safety, Vogel has picked up a scientist's train of thought and action and used it as an effective narrative device for capturing and holding a reader's attention. The subject itself is not particularly important; narrative components dictate whether Vogel writes a short story or a science story. In this case, through her understanding of both narrative structure and necessary components, she produces the latter, although depending on the medium, a writer can—and at times must—mix up the order.

This sequence of observation, question, testable assertion, test of the assertion, and action probably is the most common property of all science. Yet the general public, which doesn't habitually think like scientists, typically stops with an assertion that is either not testable or interprets observations to fit some preconceived notion of how the universe operates. The public also can easily demand an action without prior investigation or in the face of evidence that does not support the action.

Much of our decision making throughout most of our formative years is based on desire, perception, and ideology imposed on us by parents, teachers, and churches. Rare indeed is the high school student who has heard adverbs other than "should" and "must" modifying actions such as driving, spending money, studying, and preparing for examinations.

Yet those adverbs are rare in the realm of scientific inquiry unless they are part of a plan to test a claim and are negotiable as a result of observations integral to the test. In these situations, when testing assertions about natural phenomena, "must" can easily evolve into "cannot," and vice versa, when evidence counteracts perception or desire.

Whether of their own volition or from exposure to certain habits of mind, young scientists readily accept this negotiability as integral to their world view, in the process eroding, if not eliminating altogether, reliance on received wisdom.

Kristine Kelly (p. 52) illustrates this habit-of-mind issue beautifully. Her observation that "people are more comfortable in buying and selling stocks on the Internet than in talking about genetics" is actually recognition of the ongoing informal education (the "should" and "must" factors) we get on money management from the day we're first given a weekly allowance, thus producing our ability and our willingness to link certain kinds of evidence with action.

But unlike the case with genetics, the evidence surrounding financial transactions usually is very easy to gather ("I have enough cash to buy that thing"). In addition, both the motivation and outcome are clear and immediate ("I want/need that item so I'll get it"). In addition, "cost," "income," "bank account," "salary," "interest," and "monthly payments" are all vocabulary terms that would have been quite foreign to aboriginal Americans in the early 1700s but are lingua franca even for today's high school students.

Thus public education, provided largely by the media and integral to the business world, gives us language to describe and understand money; no such daily bombardment with words, ideas, and stories occurs when the subject is genetics. Kelley's point about language is a key one: Argot quickly stifles communication. "Highway crack" may be common parlance to most of us, but to Abby Vogel's Virginia DOT worker it's argot because it's data; the context of its use has changed its definition.

Boyce Rensberger's comment (p. 69) that "lay-language news-papers, magazines, and television programs form the only sig-nificant media through which adults in the general public learn of current research and about possible controversies related to it" captures the relationship between journalists and scientists, and probably between journalists and any other profession. The key term in this statement, however, is "lay-language." Science writers are translators.

Rensberger's point about specialization also is a highly rel-evant one, namely that because of the very nature of their jobs, journalists covering science cannot specialize, whereas scientists *must* specialize in order to be successful at their work.

His advice to scientists waiting for a reporter is excellent (p. 71), although a little bit idealistic: Those who would actually read this advice and take it to heart probably don't need it in the first place; those who need it the most would likely disdain it because they're terrified of reporters. Thus Rensberger's advice to journalists sent on science missions is perhaps best considered a technique for reducing this fear. A scientist's worst nightmare is being misrepresented; a few important vocabulary words, understood and used in a reporter's question, change completely the nature of any scientist-journalist encounter, usually to the benefit of the writer and the resulting story.

David Ehrenstein's point about analogy and imagery also is an excellent one (p. 75), and may provide an opportunity for scientists to learn from journalists. Scientists are constantly presenting their work to other scientists, typically believing that everyone in the audience understands perfectly why the research is being done, how it's being done, and what is being said about it.

This belief is not always very well founded; a post-doc inter-viewing for a position in academia and presenting a research seminar to the department should never assume that this group of fellow scientists has any clue about what's being said from the front of the room. Scientists today are often so specialized and the literature of their own specialty so voluminous that they rarely have or take the time to broaden their knowledge base. To be successful, therefore, our post-doc interviewee must act like a journalist, preparing the audience for the heavier stuff, such as hypotheses, data, and statistics, by drawing analogies, using plain language, and establishing context in ways appropriate to a news story.

Leslie Fink's paper (p. 6) contains a highly creative idea, passed along from a group of scientists at Cornell who were developing a graduate course on science communication. This group proposed that budding scientists engage in role-playing: visiting local news media offices and studios, sitting in on editorial meetings, doing reporter-like interviews with fellow students, and writing press releases for their own research.

Role-playing is known to be a highly effective teaching technique because it accomplishes two things: first, it evokes learning-by-doing but with few real-life consequences for doing something wrong; and second, perhaps most importantly, it changes the context within which an activity is conducted. By moving from one arena to another — from the seminar room to the newsroom — a student scientist encounters the need for transferable skills. Such encounters can validate this need in a way that no teacher can do from the front of a classroom.

Sidney Perkowitz and Eddy Von Mueller (p. 81) also address the matter of context, in this case the deliberate use of science as a character within stories whose narrative is driven by exceedingly familiar human situations, that is, the "cultural, political and textual framework" of a screenplay. The film *Armageddon* (1998) may seem heavily scientific, or at least technological, but it's actually little more than a story about heroes trying to save people from a big-time threat. This story may have originated in the Stone Age when the clan's Bruce Willis–type tough guy first stood up to a cave bear. It should be noted here that some baboons will actually attack a leopard, an ostensibly altruistic, and heroic, act on behalf of the troop; and even small birds will harass an owl, so the basic plot line for *Armageddon* might well be deeply embedded in vertebrate genes.

But Perkowitz and Von Mueller also make an exceedingly important point about the media's use of science as a character, namely that such use contributes to our attitudes toward science. Their assertion that "science fiction cinema is part of a larger societal discourse that includes scholarly journals, textbooks, journalistic treatments, and public-policy debates" is clearly validated by their comparison of two films involving climate change. They make the case that *The Day After Tomorrow* (2004, starring Dennis Quaid), set the stage (so to speak) for Al Gore's *An Inconvenient Truth* (2006). The modern narrative devices, including special effects and the film medium's often extreme compression of time, worked as a translator, relieving viewers of the "scientist's burden," namely, years of original research and acquisition of vocabulary. Such research gives an individual access to the primary literature, an access that tends to compress decades of discovery into a few short sentences, even in an individual scientist's mind.

Compression of time, space, and events also is inherent in journalism and literature, even, if not especially, the kind of literature consumed by the general public today, including movies. Although a screenplay compresses time and space, the result can be highly educational. Thus the public can "see" process, cause, and effect, and nonhuman aspects of the universe, all in the same way scientists "see" those same phenomena—with the mind's eye.

Such vision validates one's interpretation of whatever observations we make: "Sure, nature behaves that way," we conclude comfortably, whereas the unexposed are likely to say, "That would never happen." Thus studies cited by Perkowitz and Von Mueller show that The Day After Tomorrow performed such a validating role with respect to An Inconvenient Truth, increasing the public's acceptance of the Gore film as a valid representation of climate-change consequences.

At a smaller scale, scientists knew full well that human modifications of the coastline outside New Orleans were a recipe for disaster; one wonders whether the aftermath of Hurricane Katrina would have been such an inconvenient truth had there been some on-going Day After Tomorrow–type lessons for the public at large.

The writers in this volume are heavily concerned with accuracy and techniques for communicating complex stories to an audience that seems to want simplicity and expediency. There is always the possibility of bias, of course, especially in the choice of subjects. But if science reporting has a bias — or at least an aspect that is perhaps not representative of science in general as a human endeavor — it is the media's focus on application, usefulness, benefits to humanity, and so on.

Boyce Rensberger recognizes this situation indirectly with his comments about the National Science Foundation's surveys on public attitudes toward science. In contrast to what one might conclude from discussions of intelligent design versus evolution, Rensberger points out that the public in general has highly positive attitudes toward science and scientists. He cites survey results in which 89 percent of the respondents agree that "most scientists want to work on things that will make life better for the average person."

This statistic should be a journalist's dream: Focus on the applied success — scientists discover a cure for some disease. Can it be long before they will kill all agricultural pests, rid the world of malaria, convert salt water into gasoline, and preserve coral reefs for tourism-based economies in the developing world?

Yes, it will be long; many years likely will pass before some truly monumental scientific problems are solved. It is also highly likely that thousands of scientists, many of them well-paid academics in tenured positions, will be exploring nature in ways that are of absolutely no immediate, or even obvious, value to society as a whole, but are deeply rewarding to the scientists themselves.

So we might ask: What is the actual product of this grand endeavor known as "science"? I submit that the best answer is: scientists. Through research designed mostly to satisfy our curiosity about the universe, we humans are teaching ourselves how to do that research, developing technology to further the work, and eventually reaping the benefits of whatever discoveries are made.

The fraction of science that actually gets reported to the public through the media is tiny; the fraction of science that is actively engaged in producing young scientists and developing new ways of conducting science is very large. This aspect of science is what is mostly not understood by the public: You get the benefits from science only if you do a lot of it, because you can never tell when a truly useful discovery will be made by people finding something they weren't looking for in the first place.

Finally, we might ask: What really is the cost, to a nation, of a public that not only is scientifically illiterate, but also is inclined to consider science as simply another lens through which we view the world, playing the same role, more or less, as politics, religion, and economics? The answer, of course, is: plenty.

Gene Whitney (p. 21) explains why this answer is so, although his advice for reducing that cost ("Steps for Informing Policy") can seem outright subversive to scientists who routinely experience serious peer review and whose work can be highly sequestered and focused for years on end. Whitney's version of a twelve-step program is right on target, although, perhaps idealistically, he addresses scientists who "are excellent communicators" and are interested in the policymaking process.

The take-home message, not only from Whitney's paper but also from all the contributors to this volume, is that whether we believe it or not, every citizen has a vital interest in policy that addresses questions involving science, technology, and the natural world, and that being "excellent communicators" must be a goal for scientists in general, not just a lucky toss of the genetic dice for a few.